T0213577

Lecture Notes in Business Information Processing   **300**

Series Editors

Wil M.P. van der Aalst
  Eindhoven Technical University, Eindhoven, The Netherlands
John Mylopoulos
  University of Trento, Trento, Italy
Michael Rosemann
  Queensland University of Technology, Brisbane, QLD, Australia
Michael J. Shaw
  University of Illinois, Urbana-Champaign, IL, USA
Clemens Szyperski
  Microsoft Research, Redmond, WA, USA

More information about this series at http://www.springer.com/series/7911

Stanisław Wrycza · Jacek Maślankowski (Eds.)

Information Systems: Research, Development, Applications, Education

10th SIGSAND/PLAIS EuroSymposium 2017
Gdansk, Poland, September 22, 2017
Proceedings

 Springer

Editors
Stanisław Wrycza
Department of Business Informatics
University of Gdansk
Sopot
Poland

Jacek Maślankowski
Department of Business Informatics
University of Gdansk
Sopot
Poland

ISSN 1865-1348   ISSN 1865-1356   (electronic)
Lecture Notes in Business Information Processing
ISBN 978-3-319-66995-3   ISBN 978-3-319-66996-0   (eBook)
DOI 10.1007/978-3-319-66996-0

Library of Congress Control Number: 2017952381

Printed on acid-free paper

This Springer imprint is published by Springer Nature
The registered company is Springer International Publishing AG
The registered company address is: Gewerbestrasse 11, 6330 Cham, Switzerland

Preface

Systems analysis and design (SAND), or Business Informatics, as it is more frequently called in Europe, constitutes the classical field of research and education in the area of management information systems (MIS). SAND continuously attracts the attention of both academia and business. The rapid progress made in ICT naturally generates the requirements for the new generation of methods, techniques, and tools of SAND, adequate for modern IS challenges. Therefore, international thematic conferences and symposia have become widely accepted forums for the exchange of concepts, solutions, and experiences in SAND. In particular, the Association for Information Systems (AIS) is undertaking a number of initiatives towards SAND's international development.

The objective of the EuroSymposium on Systems Analysis and Design is to promote and develop high-quality research on all issues related to SAND. It provides a forum for SAND researchers and practitioners in Europe and beyond to interact, collaborate, and develop their field. The Eurosymposia were initiated by Prof. Keng Siau as the SIGSAND – Europe Initiative. Previous EuroSymposia were held at:

- University of Galway, Ireland – 2006
- University of Gdansk, Poland – 2007
- University of Marburg, Germany – 2008
- University of Gdansk, Poland – 2011
- University of Gdansk, Poland – 2012
- University of Gdansk, Poland – 2013
- University of Gdansk, Poland – 2014
- University of Gdansk, Poland – 2015
- University of Gdansk, Poland – 2016

The accepted papers of the EuroSymposia held in Gdansk were published as follows:

- 2nd EuroSymposium 2007: A. Bajaj, S. Wrycza (eds), Systems Analysis and Design for Advanced Modeling Methods: Best Practises, Information Science Reference, IGI Global, Hershey, New York, 2009
- 4th EuroSymposium 2011: S. Wrycza (ed.) 2011, Research in Systems Analysis and Design: Models and Methods, series: LNBIP 93, Springer, Berlin 2011
- Joint Working Conferences EMMSAD/EuroSymposium 2012 held at CAiSE'12: I. Bider, T. Halpin, J. Krogstie, S. Nurcan, E. Proper, R. Schmidt, P. Soffer, S. Wrycza (eds.) 2012, Enterprise, Business-Process and Information Systems Modeling, series: LNBIP 113, Springer, Berlin 2012
- 6th SIGSAND/PLAIS EuroSymposium 2013: S. Wrycza (ed.), Information Systems: Development, Learning, Security, Series: Lecture Notes in Business Information Processing 161, Springer, Berlin 2013

- 7th SIGSAND/PLAIS EuroSymposium 2014: S. Wrycza (ed.), Information Systems: Education, Applications, Research, Series: Lecture Notes in Business Information Processing 193, Springer, Berlin 2014
- 8th SIGSAND/PLAIS EuroSymposium 2015: S. Wrycza (ed.), Information Systems: Development, Applications, Education, Series: Lecture Notes in Business Information Processing 232, Springer, Berlin 2015
- 9th SIGSAND/PLAIS EuroSymposium 2016: S. Wrycza (ed.), Information Systems: Development, Research, Applications, Education, Series: Lecture Notes in Business Information Processing 264, Springer, Berlin 2016

There are three organizers of the 10th EuroSymposium on Systems Analysis and Design:

- SIGSAND – Special Interest Group on Systems Analysis and Design of AIS
- PLAIS – Polish Chapter of AIS
- Department of Business Informatics of University of Gdansk, Poland

SIGSAND is one of the most active SIGs with quite a substantial record of contributions for AIS. It provides services such as the annual American and European Symposia on SIGSAND, research and teaching tracks at major IS conferences, listserv, and special issues in journals.

The Polish Chapter of the Association for Information Systems (PLAIS) was established in 2006 as the joint initiative of Prof. Claudia Loebbecke, former President of AIS, and Prof. Stanislaw Wrycza, University of Gdansk, Poland. PLAIS co-organizes international and domestic IS conferences.

The department of Business Informatics of the University of Gdansk is conducting intensive teaching and research activities. Some of its academic manuals are bestsellers in Poland. The department is active internationally. The most significant conferences organized by the department were: the 10th European Conference on Information Systems, ECIS 2002, and the International Conference on Business Informatics Research, BIR 2008. The department is a partner of the ERCIS consortium – European Research Center for Information Systems.

EuroSymposium 2017 had an acceptance rate of 22%, with submissions divided into the following three groups:

- Data Analytics
- Web-Based Information Systems
- Information Systems Development

The accepted papers reflect the current trends in the field of systems analysis and design.

I would like to express my thanks to all authors, reviewers, advisory board members, and the members of the International Program Committee and Organizational Committee for their support, efforts, and time. They have made possible the successful accomplishment of EuroSymposium 2017. The 10th SIGSAND/PLAIS EuroSymposium is organized in conjunction with the Golden Jubilee (50 years) of Business Informatics at the University of Gdansk.

July 2017   Stanislaw Wrycza

Organization

General Chair

Stanislaw Wrycza   University of Gdansk, Poland

Organizers

- SIGSAND - the Association for Information Systems (AIS) Special Interest Group on Systems Analysis and Design
- The Polish Chapter of the Association for Information Systems - PLAIS
- The Department of Business Informatics at the University of Gdansk

Advisory Board

Wil van der Aalst	Eindhoven University of Technology, The Netherlands
David Avison	ESSEC Business School, France
Joerg Becker	European Research Centre for Information Systems, Germany
Jane Fedorowicz	Bentley University, USA
Alan Hevner	University of South Florida, USA
Dimitris Karagiannis	University of Vienna, Austria
Helmut Krcmar	Technical University of Munich, SAP University Competence Center, Germany
Claudia Loebbecke	University of Cologne, Germany
Keng Siau	Missouri University of Science and Technology, USA
Roman Słowiński	Committee on Informatics, Polish Academy of Sciences, Poland

International Program Committee

Stanislaw Wrycza	University of Gdansk, Poland
Özlem Albayrak	Bilkent University, Turkey
Eduard Babkin	Higher School of Economics, Moscow, Russia
Akhilesh Bajaj	University of Tulsa, USA
Palash Bera	Saint Louis University, USA
Lasse Berntzen	University College of Southeast, Norway
Sven Carlsson	Lund University, Sweden
Witold Chmielarz	University of Warsaw, Poland
Petr Doucek	University of Economics, Prague, Czech Republic

Reima Suomi	University of Turku, Finland
Jakub Swacha	University of Szczecin, Poland
Pere Tumbas	University of Novi Sad, Serbia
Catalin Vrabie	National University, Romania
Yinglin Wang	Shanghai University of Finance and Economics, China
H. Roland Weistroffer	Virginia Commonwealth University, USA
Carson Woo	Sauder School of Business, Canada
Karolina Zmitrowicz	WSB University in Gdansk, Poland
Iryna Zolotaryova	Kharkiv National University of Economics, Ukraine
Joze Zupancic	University of Maribor, Slovenia

Organizing Committee

Chair

Stanislaw Wrycza

Secretary

Anna Węsierska	Department of Business Informatics at University of Gdansk

Members

Dorota Buchnowska	Department of Business Informatics at University of Gdansk
Bartłomiej Gawin	Department of Business Informatics at University of Gdansk
Przemyslaw Jatkiewicz	Department of Business Informatics at University of Gdansk
Dariusz Kralewski	Department of Business Informatics at University of Gdansk
Michał Kuciapski	Department of Business Informatics at University of Gdansk
Bartosz Marcinkowski	Department of Business Informatics at University of Gdansk
Jacek Maslankowski	Department of Business Informatics at University of Gdansk

EuroSymposium 2017 Topics

Agile Methods
Big Data
Business Process Modeling
Cloud Computing
Conceptual Modeling
Cognitive Issues in SAND

Crowdsourcing and Crowdfunding Models
Design of Mobile Applications
Design Theory
Case Studies in IS
Enterprise Architecture
Enterprise Social Networks
Ethical and Human Aspects of IS Development
Human-Computer Interaction
Information Systems Development: Methods and Techniques
Internet of Things
Model-Driven Architecture
Ontological Foundations of SAND
Open Source Software (OSS) Solutions
Project Management
Requirements Engineering
Research Methodologies in IS
SAND in ERP and CRM Systems
SCRUM Approach
Smart Information Systems
Social Media Analytics
Social Media Use in Business
Social Networking Services
Software Intensive Systems and Services
Teaching SAND
Teams and Teamwork in IS
Towards Data Driven Information Systems
User Experience (UX) Design
UML, SysML, BPMN
Workflow Management

Contents

Data Analytics

Data Cleaning Technique for Security Logs Based on Fellegi-Sunter Theory

Diana Martinez-Mosquera[1]([✉]) [iD], Sergio Luján-Mora[2] [iD],
Gabriel López[3] [iD], and Lauro Santos[4] [iD]

[1] Departamento de Ciencias de la Ingeniería, Universidad Israel, Quito, Ecuador
smartinez@uisrael.edu.ec
[2] Department of Software and Computing Systems, University of Alicante,
Alicante, Spain
slujan@ua.es
[3] Departamento de Electrónica, Telecomunicaciones y Redes de Información,
Escuela Politécnica Nacional, Quito, Ecuador
gabriel.lopez@epn.edu.ec
[4] Performance Testing and Continuous Integration, Nokia Solutions
and Networks, Amadora, Portugal
lauro.santos@nokia.com

Abstract. Information security is one of the most important aspects an organization should consider. Due to this matter and the variety of existing vul-
nerabilities, there are specialized groups known as Computer Security Incident
Response Team (CSIRT), that are responsible for event monitoring and for
providing proactive and reactive support related to incidents. Using as a case
study a CSIRT of a university with 10,000 users, and considering the high
volume of events to be analyzed on a daily basis, it is proposed to implement a
Big Data ecosystem. One of the most important activities for the information
processing is the data cleaning phase, it will remove useless data and help to
overcome storage limitations; since CSIRT is actually limited to a small
time-frame, usually a few days and cannot analyze historical security events.
Focusing on this cleaning phase, this article analyzes an intuitive technique and
proposes a comparative technique based on the Fellegi-Sunter theory. The main
conclusion of our research is that some data could be safely ignored helping to
reduce storage size requirements. Moreover, increasing the data retention will
enable to detect some events from historical data.

Keywords: Data · Cleaning · Big Data · Security · Fellegi-Sunter

1 Introduction

Big Data is a concept related to big volumes of information, which complies with
attributes like Volume, Velocity, Variety, Variability, Viscosity, Virality, Veracity,
Viability, Visualization, and Value [1]. Security events can be considered as Big Data
due to the high volume of information generated by the data network events, which
requires fast analysis, and depending on the type of technology, data can be structured,
semi structured, or unstructured. Furthermore, veracity of data is accomplished since

© Springer International Publishing AG 2017
S. Wrycza and J. Maślankowski (Eds.): SIGSAND/PLAIS 2017, LNBIP 300, pp. 3–12, 2017.
DOI: 10.1007/978-3-319-66996-0_1

data comes from trusted equipment in an advisable form, which has a high value for the organization, and finally has to be visualized for its respectively analysis [2].

Management of security alerts requires considerable storage capacity since companies generate between 10 and 100 billion security events every day [3]. Therefore, analyzing existent data cleaning techniques is important and how to adapt them to a Big Data ecosystem is an open issue.

This paper takes as case study a Computer Security Incident Response Team (CSIRT) of a university with 10,000 users, responsible for analyzing all the security events in the network; for this proposal, a Big Data ecosystem have been implemented in a testbed [4]. The research is focused primarily on the cleaning phase, due to the CSIRT's space storage constraints and since they cannot currently analyze historical security events; for such goal, an intuitive technique is analyzed and a comparative technique based on Fellegi-Sunter theory [5] is proposed.

The intuitive technique accomplishes to discard irrelevant data from security logs, which has to be identified by organization expert personnel in security information; the main issue with this technique is the requirement of human intervention [4]. Therefore, this paper proposes a comparative technique to automatically identify which data fields should be discarded from the security logs based on the Fellegi-Sunter theory and the Levenshtein Distance [5–7].

The comparative data cleaning solution has been tested with real data, and the main objective is to reduce the size of the data before their analysis. One of the most important challenges of this investigation is to adapt automatic data cleaning techniques to a Big Data ecosystem. The following questions are proposed to be solved:

- Is it possible to adapt an automatic data cleaning technique to a Big Data ecosystem?
- Is it possible to reduce the size of the security data without affecting security breaches detection by implementing a comparative technique based on the Fellegi-Sunter theory?

This paper is organized with the following structure; Sect. 2 shows the existing related work; Sect. 3 analyzes an intuitive data cleaning technique; Sect. 4 presents the proposed comparative data cleaning technique based on the Fellegi-Sunter theory; Sect. 5 shows the results of the comparative technique; and conclusions and future work are shown in Sect. 6.

2   Related Work

Data cleaning techniques allow the removal of irrelevant or unnecessary items in the analyzed data [8]. The following presents relevant work related to the data cleaning process:

- T. Aye has mentioned the importance of web data preprocessing in order to improve mining process efficiency, and his study has considered irrelevant data pruning as a main task, therefore it is the main research goal of the investigation to apply this affirmation focused on security logs [8].
- Maletic and Marcus evidenced that real implementations do use customized data cleaning processes and, in conclusion, have shown the need for building higher quality tools [9].
- Khayyat et al. described an efficient technique called Big Dansing targeted to Big Data; their main purpose is to remove inconsistencies on stored data, while our goal is to clean the data before storing [10].
- Krishnan et al. proposed the idea of making an easy and fully automated data cleaning process to avoid the existence of human intervention, and it has been some of the motivation to our research [11].
- Winkler explained how the Fellegi-Sunter mathematical model does not need trained data to find duplicated data within files or group of files, thus, our proposed comparative technique uses such theory [12].

Following Krishnan et al. [11], and in order to avoid the existence of human intervention, this paper proposes the use of the Fellegi-Sunter mathematical model to perform an automated data cleaning of security logs by using a real network as case study.

3 Intuitive Method for Security Data Cleaning

This intuitive method requires human intervention to determine the irrelevant security data: Information security experts analyze available information and then determine which fields can be discarded.

3.1 Log File Structure

CSIRT's firewall writes around 1,000,000 lines per hour in the log about the network events, this type of information is semi structured and the different fields can be easily identified, since they are comma-delimited. Figure 1 shows 77 header fields available on the firewall log and Fig. 2. presents the resulting 26 header fields after applying the intuitive method for data cleaning. Here, CSIRT's security experts analyze the firewall log file and decide which fields can be discarded, according their report requirements to detect a strange behavior in the network.

An important question to be considered is the expensive cost associated to customize the CSIRT's firewall configuration for discarding the fields directly in the log generation, since it is a commercial solution.

Time, Description, Interface Name, Interface Direction, Interface, Action, Destination, Inzone, NAT Additional Rule Number, NAT Rule Number, Origin, Out-Zone, Policy Date, Policy Management, Policy Name, Blade, Product Family, Protocol, Rule, Rule Name, Rule UID, Source Port, Destination Port, Service ID, Service Name, Source, Type, XlateDPort, XlateDst, XlateSport, XlateSrc, Matched Category, Application Description, Application ID, Application Properties, Application Risk, Application Rule ID, Application Rule Name, Application Signature ID, Application Name, Primary Category, Proxy Source IP, Browse Time, Total Bytes, Received Bytes, Sent Bytes, Suppressed Logs, referrer_self_uid, Resource, Client Type, Attack Name, Attack Information, Confidence Level, Industry Reference, Performance Impact, Protection Name, Protection Type, Protection ID, Severity, IPS Profile, Update Version, UserCheck ID, ICMP, ICMP Code, ICMP Type, Message Information, DLP Incident UID, Frequency, Log ID, UserCheck Message to User, Confirmation Scope, Total Logs, Server Type, reason, Packet Information, Authentication Method, Authentication Status, authentication trial, Client Name, Product Version, Description, Endpoint IP, Identity Source, Identity Type, Session ID, Source User Name, User, referrer_parent_uid, Captured UUID, tcp flags, Duration, Termination Reason, Firewall Message, Session Identification, Connection UID, IKE Initiator Cookie, IKE Responder Cookie, Destination Key ID, ike ids:, ike:, Encryption Methods, IKE Phase2 Message ID, VPN Peer Gateway, Encryption Scheme, Source Key ID, Start Time, VPN Feature, Content Type, Destination Phone Number, Source IP-phone, VoIP Call ID, VoIP Configuration, VoIP Log Type, Request, VoIP Reject Reason Information, VoIP Reject Reason, streaming engine, Malware Action, Malware Family, Malware Rule ID, packet_capture_name, packet_capture_time, packet_capture_unique_id, remediation_options, Scope, Reject ID, encryption failure:, log id, Session UID, Resource Probing, Comment, Status, update_service, Version, CPU Utilization Percent, Memory Utilization Percent, Resource Shortage, Duration in Seconds, Fragments Dropped, IP ID, IP Length, IP Offset, Message, Event Count, summary, sys_message:, contract_name, Special properties, Subscription Expiration, subscription_stat, subscription_stat_desc, stormagentaction, stormagentname, Authentication Encryption Methods, Operating System, Product Build number, Mobile Access Category, Device ID, Event Type, Host IP, Host Type, Host Name, Login Timestamp, MAC Address, Office Mode IP, OS Bits, OS Build, OS Name, OS Version, Re-authentication every, suppressed_logs, Data Protocol, User DN, User Group, Source User Group, blade_name, Information

Fig. 1. Security log file headers.

Time, Description, Interface Name, Interface, Action, Destination, Inzone, Origin, Protocol, Rule, Rule Name, Source Port, Destination Port, Service ID, Service Name, Source, Type, XlateDPort, XlateDst, XlateSport, XlateSrc, Matched Category, Application Description, Application Name, Proxy Source IP, Resource

Fig. 2. Useful security log file headers.

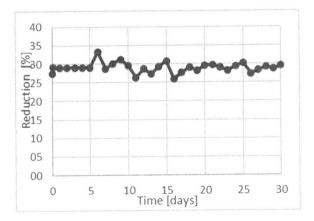

Fig. 3. Intuitive method results [4]

3.2   Intuitive Method Results

After applying the intuitive method for CSIRT's firewall data, relevant size reductions between 25% and 30% can be achieved, whereas the attained average is 28.9% as shown in Fig. 3; there x axis belongs to the data time and y axis to the accumulated reduction. With it, storage costs can be reduced or data retention capability increased by at least 25% [4]. However, the main problem is the need of expert intervention to identify the irrelevant data, while the data cleaning process is carried out, for instance, whether the firewall configuration is changed the intuitive method has to be applied again. Thus, this paper proposes an automatic method to remove human intervention from the data cleaning process.

4   Comparative Proposal Based on Fellegi-Sunter Theory

Data cleaning rules vary from organization to organization and, according to their requirements, they can discard or not a field from a security log. Thus, our research proposes a technique for, based on the user final report, generating these cleaning rules in an automatic manner.

Our main goal is to avoid the human intervention in the data cleaning process; thus, the irrelevant data could be identified by comparing two files, the original security log file and the final user security report. This last one will have the useful data that must be kept from the original file; for instance, for horizontal, vertical and box scanning reports, the useful fields are: Source and target IP addresses, ports, domains, and time.

In order to achieve the research main goal, we selected the Fellegi-Sunter model [5] since it describes a mathematical theory for record linkage without the need of data training. The model offers a framework for solutions focused on record recognition from two files that can represent people, objects or events, the last one being our point of interest [5].

Fellegi and Sunter defined three sets of elements, A1, A2 and A3, where A1 corresponds to a match between two files; A3 corresponds to a non-match and A2 to a possible match [12]. We can base our decisions in A1 and A3, where A1 corresponds to a match between A and B files and A3 corresponds to a non-match or non-link between A and B files. As A, we will denote the set of fields in the final user report; this report will be obtained without the data cleaning process. As B, we will denote the set of fields of the log file gathered from the CSIRT's firewall.

We have worked to implement this comparative technique and present the attained results. The main challenge was to compare the original log file and the final user report, since they have header field names in different languages; the first one with English headers as it can be observed in Fig. 4, and resulting reports, with Spanish headers as it is shown in Fig. 5. It caused header fields not to be considered but instead only their data fields.

```
Time,Description,Interface Name,Interface Direction,Interface,Action,Destination
"01/Mar/2017,08:04:43",Accepted on rule 427 (Acceso Navegaci n),eth2-01,outbound
"01/Mar/2017,08:09:54",Accepted on rule 427 (Acceso Navegaci n),eth2-01,outbound
```

Fig. 4. Original log fragment.

Enlazado con Resumen de eventos								
Grave....	Mensaje de regla		Recu....	IP de origen	IP de destino	Protocolo	Última vez	Subtipo d

Fig. 5. Final user report headers.

The main steps to achieve this comparison were identifying the data type in the user report, performing a match with the original log file and comparing the fields. Below a detailed explanation.

The initial data to be processed:

A   Set of fields of the final user report.
$\propto(a)$ Elements of A.
B   Set of fields of the original log file.
$\propto(b)$ Elements of B.

A and B elements must be compared, here we have experimented comparing the fields row by row, since, as we mentioned before, the header fields are not in the same language.

$$\propto(a) \times \propto(b) \tag{1}$$

After the comparison we obtained subset M as result and we denote $\propto(m)$ as the elements of M.

$\propto(m)$ Elements of B that match elements of A

Algorithm 1. Comparative Technique

```
Input: File A
Output: C and D files
Begin
Read every comma-delimited field, from the second row of file
A and identify:
    If field = number do;
        Send true value to function KEEP NUMBERS in file C;
    else
    If field = string do;
            If field = date do;
            Send true value to KEEP DATES function in file C;
            else
            If field = IP address do;
            Send true value to KEEP IP ADDRESSES function in
            file C;
            else
            If field contains a protocol name do;
            Send true value to KEEP PROTOCOL NAMES function in
            file C;
            else
            Find the string in a separated file D;
            If the string does exist do;
            end
            else
            Copy the string in file D;
    else
    end.

/*If the functions KEEP NUMBERS, KEEP DATES, KEEP IP
ADDRESSES and KEEP PROTOCOL NAMES are already set as true in
file C, keep them*/
```

The procedure to compare the files A and B is described in the following steps:

1. Copy and number all header fields from file B in a separate file H.
2. Determine the type of the elements from file A using the Algorithm 1.
3. Read functions file C and identify the corresponding header fields in file B related to numbers, dates, IP addresses and protocols with intent to copy the header field numbers from file H into file F.
4. To identify the B header fields for the strings in file D, we propose the use of the Edition Distance or Levenshtein Distance (LD) [13]. The LD between B and D files strings (except the fields that have been identified as numbers, dates, IP addresses, or protocols), will permit to define the minimum number of editing operations needed for transforming B strings into D strings. Two strings are considered similar if the LD is lower or equal to a predefined threshold, for this case, since the strings have to be similar, the threshold is equal to 0.3 [6, 7].

5. Once the strings are identified in file B, the header field numbers corresponding to these strings must be identified in the file H and then written into file F. It is recommended to avoid repeating the header field numbers in file F.
6. Finally, we can delete the columns from file B not matching the header numbers in file F in order to obtain the file M.

5   Results

To test our comparative proposal applied to security logs cleaning, we collected data from a real network. Figure 1 can be referred for a list of available headers present on collected data and Fig. 2. to the list of possible header fields present on user reports.

For this cleaning process, first, we identified the type of the fields in the final user report, thus, numbers, dates, IP addresses and protocols were found and written to respective file C, for then identify the corresponding header numbers from file H and write them into file F. Moreover, we had to compare around 300 strings between files D and B through Levenshtein Distance.

As example, Table 1 presents the LD comparison for the second row of the D (horizontal) and B (vertical) files. There we can see that when two strings are matched the LD is equal to 0, moreover, it is important to keep the string format to avoid problems with the comparison, for instance, ï¿½. Some fields have been hidden since they contain confidential information.

Table 1. Levenshtein Distance for the second row in B file.

Fields	Accepted on rule 426 (Acceso Navegaciï¿½n)	inbound	eth2-01 inbound	Accept	Network	Log
Accepted on rule 426 (Acceso Navegaciï¿½n)	0	41	41	35	41	41
eth2-01	41	7	8	7	7	7
inbound	41	0	15	7	7	7
eth2-01 inbound	41	15	0	15	15	15
Accept	35	7	15	0	7	6
Internal	41	5	15	8	8	8
External	41	7	15	7	7	8
Today 09:35:39	41	25	25	25	25	24
Standard	41	8	14	8	8	8
Firewall	41	8	15	7	8	8
Network	41	7	15	7	0	7
Acceso Navegaciï¿½n	37	19	19	2	19	19
Log	41	7	15	6	7	0

Time, Description, Interface Name, Interface, Action, Destination, Inzone, NAT Additional Rule Number, NAT Rule Number, Origin, Protocol, Rule, Rule Name, Source Port, Destination Port, Service ID, Service Name, Source, Type, XlateDPort, XlateDst, XlateSport, XlateSrc, Matched Category, Application Description, Application ID, Application Name, Proxy Source IP, Browse Time, Total Bytes, Received Bytes, Sent Bytes, Suppressed Logs, Resource, Update Version.

Fig. 6. File F content, filled with the useful security data headers.

Taking Table 1 as reference, we selected the fields where LD < threshold and wrote the corresponding header numbers from file H into file F. At the end of this process, file F will hold the 35 headers presented in Fig. 6, which correspond to relevant data that should be kept in the log to obtain the file M.

While it was possible to discard 51 fields using the intuitive method and only 42 fields through the comparative method, the file size reduction was possible and, for this example, a security log file of 526 MB was reduced to 248 MB without any human intervention.

Comparing our research results with other researches, the main difference is the nature of the data, since we consider all security rows as equally important, while other authors propose to discard data at row level [9].

6   Conclusions and Future Work

Data cleaning is an important process to discard irrelevant data in order to prepare relevant data for other processes such as data mining, statistical analysis, etc.

This paper presents an intuitive method for data cleaning and proposes a comparative method in order to avoid the human intervention, when determining what fields can be discarded from a security log file before its analysis. The proposed technique is based on Fellegi-Sunter theory to compare original log files and final user reports, and Levenshtein Distance to identify similar strings. We presented a security log as example, but it can be applied to other logs with similar format like network logs.

Comparing the intuitive and comparative methods, the first technique allows better file size reductions; however, the proposed automatic technique also offers file size reductions with no human intervention. For the provided example in Table 1 size reductions around 47% and 46.5% can be achieved, with intuitive and comparative methods respectively.

As future work, we recommend algorithm implementation in real scenario to assess the processing time and resources requirements, and the usage of machine learning techniques for improving data cleaning processes focused on security log files.

Acknowledgements. We thank to the National Polytechnic School CSIRT for their collaboration and facilities needed to test this data cleaning technique.

References

1. Qaiyum, S., Aziz, I.A., Jaafar, J.B.: Analysis of Big Data and quality-of-experience in high-density wireless network. In: 2016 3rd International Conference on Computer and Information Sciences (ICCOINS), pp. 287–292 (2016). doi:10.1109/ICCOINS.2016.7783229
2. Arputhamary, B., Arockiam, L.: Data integration in Big Data environment. Bonfring Int. J. Data Mining **5**(1), 1–5 (2015). doi:10.9756/BIJDM.8001
3. Cárdenas, A., Manadhata, P., Rajan, S.: Big Data analytics for security. IEEE Secur. Priv. **11**(6), 74–76 (2015). doi:10.1109/MSP.2013.138
4. Martínez-Mosquera, D., Luján-Mora, S.: Data cleaning technique for security Big Data ecosystem. In: Proceedings of the 2nd International Conference on Internet of Things, Big Data and Security, vol. 1, pp. 380–385 (2017). doi:10.5220/0006360603800385
5. Fellegi, I.P., Sunter, A.B.: A theory for record linkage. J. Am. Stat. Assoc. **64**(328), 1183–1210 (1969). doi:10.1080/01621459.1969.10501049
6. Luján Mora, S., Palomar Sanz, M.: Reducing inconsistency in integrating data from different sources. In: Proceedings 2001 International Database Engineering and Applications Symposium (IDEAS 2001), pp. 209–218 (2001). doi:10.1109/IDEAS.2001.938087
7. Luján Mora, S., Palomar Sanz, M.: Comparing string similarity measures for reducing inconsistency in integrating data from different sources. In: Proceedings of the Second International Conference in Advances in Web-Age Information Management (WAIM 2001), pp. 191–202 (2001). doi:10.1007/3-540-47714-4_18
8. Aye, T.T.: Web log cleaning for mining of web usage patterns. In: 2011 3rd International Conference Computer Research and Development (ICCRD), vol. 2, pp. 490–494 (2011). doi:10.1109/ICCRD.2011.5764181
9. Maletic, J.I., Marcus, A.: Data cleansing: a prelude to knowledge discovery. In: Maimon, O., Rokach, L. (eds.) Data Mining and Knowledge Discovery Handbook, pp. 19–32. Springer, USA (2009). doi:10.1007/978-0-387-09823-4_2
10. Khayyat, Z., Ilyas, I.F., Jindal, A., Madden, S., Ouzzani, M., Papotti, P., Yin, S.: Bigdansing: a system for Big Data cleansing. In: ACM SIGMOD International Conference on Management of Data, pp. 1215–1230 (2015). doi:10.1145/2723372.2747646
11. Krishnan, S., Haas, D., Franklin, M., Wu, E.: Towards reliable interactive data cleaning: a user survey and recommendations. In: ACM SIGMOD/PODS Conference Workshop on Human. In the Loop Data Analytics (2016), p. 9. doi:10.1145/2939502.2939511
12. Winkler, W.E.: Using the EM algorithm for weight computation in the fellegi-sunter model of record linkage. In: Proceedings of the Section on Survey Research Methods, American Statistical Association, vol. 667, p. 671 (1988)
13. Levenshtein, V.I.: Binary codes capable of correcting deletions, insertions, and reversals. Soviet Physics Doklady **10**(8), 707–710 (1966)

Understanding Benefits and Limitations of Unstructured Data Collection for Repurposing Organizational Data

Arturo Castellanos[1(✉)], Alfred Castillo[2], Roman Lukyanenko[3], and Monica Chiarini Tremblay[4]

[1] Baruch College (CUNY), New York City, NY, USA
arturo.castellano@baruch.cuny.edu
[2] Cal Poly, San Luis Obispo, USA
acast084@fiu.edu
[3] University of Saskatchewan, Saskatoon, Saskatchewan, Canada
lukyanenko@edwards.usask.ca
[4] College of William and Mary, Williamsburg, VA, USA
tremblay@fiu.edu

Abstract. With the growth of machine learning and other computationally intensive techniques for analyzing data, new opportunities emerge to repurpose organizational information sources. In this study, we explore the effectiveness of *unstructured data entry* formats in repurposing organizational data in solving new tasks and drawing novel business insights. Unstructured data accounts for more than 80% of the organizational data. Our research analyzes the implications of using unstructured data entry formats for propagation of organizational styles. We study this phenomenon in the context of case management in foster care. Using natural language processing and machine learning, we show that unstructured data formats foster entrenchment and propagation of individual organizational styles and deviations from the industry norms. Our findings have important implications both to theory and practice of business analytics, conceptual modeling, organizational theory and general data management.

Keywords: Systems analysis and design · Text mining · Stylometry · Unstructured data · Institutional theory · Case management

1 Introduction

Organizational data becomes a strategic resource for organizations. Effectively, these data can be aggregated to provide trends, plan, improve processes, support decision-making, or solve additional tasks by repurposing it. While some of these data are in structured and consistent form, organizational reports are often in unstructured format. IDC estimates that more than 80% of the enterprise data generated is unstructured [1].

Here, we define *unstructured data* as any document - clinical documentation, personal message, progress note, business report - that comprises primarily of unstructured text – with little or no predefined structure or meta data describing the content of the document. It is common to contrast unstructured data with *structured data* - such as information

S. Wrycza and J. Maślankowski (Eds.): SIGSAND/PLAIS 2017, LNBIP 300, pp. 13–24, 2017.
DOI: 10.1007/978-3-319-66996-0_2

stored in a spreadsheet or a database that follows a predefined structure or contains metadata describing the content of the stored information. Naturally, unstructured content does have internal structure, but it's semantics needs to be discovered through additional processing (e.g., natural language processing) by a computer.

Despite the pervasiveness of unstructured data in organizations, traditional IS research offers limited guidance in understanding the implications of unstructured data-entry formats in decision-making – the alignment between the information needs of data consumers and that of data contributors. Data-entry refers to how these data is entered into a system (e.g., forms, templates or free-text fields).

One of the challenges of unstructured data formats is the inherent flexibility it gives to users when entering data into an information system—this may partially explain its popularity among data users. Users may deviate from the deep structure ("the meaning") of the system by capturing different information in a field that was not originally intended for [2–5]. For example, in a study of an electronic patient record, physicians complained that the system was too "rigid" to capture the core reason of the patient's visit. To overcome this perceived limitation physicians started to use a text field labeled as "conclusion" to enter such information and regarded it as a central field for subsequent patient's visits [6, 7]. This is consistent with recent findings from the context of social media that suggest that imposing rigid structure when collecting information may result in users attempting to circumvent the structure by guessing or abandoning data entry entirely [8–11].

Despite the obvious benefits of unstructured information collection and its growing prevalence for organizational data capture and in social media environments, traditional research on conceptual modeling, systems analysis and information use mainly examined information collection in structured settings [12–14]. This creates a major gap in understanding of the limitations and benefits of the unstructured data collection, the gap where attempting to address in this and future work.

A better understanding of unstructured data collection is becoming increasingly important. Among other factors motivating our work is the on-going practice whereby organizations are repurposing data for business insight. This is possible due to increasing computational power and the availability of sophisticated analytical tools. For example, Tremblay, Berndt, Luther, Foulis and French [15] analyzed unstructured progress notes to predict falls in the elderly. Sørlie, Perou, Tibshirani, Aas, Geisler, Johnsen, Hastie, Eisen, van de Rijn and Jeffrey [16] classified breast carcinomas based on variations in gene expression patterns and then correlate tumor characteristics to clinical outcome. Larsen and Bong [17] identified intellectual communities in the field of information systems and detected discordant naming practices of constructs (e.g., same term to refer to different phenomena or using different terms to refer to the same phenomena).

We focus on the effect of inferential utility in repurposing data. Our premise is that as people specialize they are more comfortable using domain-specific language. We demonstrate the relationship between inferential utility when repurposing unstructured electronic documentation and how institutionalization of practices need to be accounted for when designing more effective systems. Our goal is to demonstrate the implications of *unstructured data entry* on the ability of organizations to repurpose their existing tactical reports for strategic insight. We study this phenomenon in the context of case management in foster care (e.g., identifying cases of psychotropic drug use).

2   SafeKids

Our research is based on the triangulation of qualitative and quantitative evidence. Specifically, we draw on own experiences with the case of foster care in the United States. This setting allowed us to examine the issues related to unstructured data formats in a concrete and real scenario. This enabled us to produce qualitative insights into the nature of organizational reporting and the role of data formats. At the same time, we undertook systematic data collection from the foster care organizations we were observing to provide systematic analysis of the data and draw statistical inferences. We then returned to our qualitative understanding of the setting for additional corroboration and support of the quantitative evidence provided.

The organization that supplied the data for this paper is called SafeKids (name is anonymized) – an American non-profit corporation created by advocacy communities to oversee several Full-Case Management Agencies (FCMAs) that provide full case management services. Many of these cases include children at-risk of abuse and/or neglect. Failure to identify at-risk clients is highly problematic, because adverse outcomes can include serious adverse events—including death. Since data are often encoded in free-text form (e.g., reports, encounter notes, case notes, progress notes), we study the impact of different data-entry formats, in particular, when the goal is to repurpose these notes and use them for solving a different tactical need. We do so with a case study in which the tactical need is to identify children that are taking psychotropic medicines by analyzing the child's case notes—as reported by caseworkers when visiting their homes.

In previous research Castillo et al. [18] hypothesized that by using these home-visit notes, which contained the child's record and behavior (e.g. has signs of abuse and neglect, aggressive behavior), they could identify children taking psychotropic medication by training Statistical Text Mining (STM) classification models [19]. An interesting result was that models trained on individual FCMAs data had varying levels of classifications accuracy. This led us to ponder, if all agencies are not equal, did the writing style of each FCMA have an effect in improving the accuracy of our classification model? We turn to organizational theory and psychology theory to understand the underpinnings of flexibility in data-entry tasks.

3   Proposition Development

Institutions are organized and established by procedures that guide the actions of individuals [20]. Organizational activity (social and non-social) can become a pattern that is repeated by individuals in the organization. The rules, norms, and meanings arise in interaction and are preserved and modified by the behavior of individuals over time [21, 22]. In the absence of contextual change, actors are more likely to replicate scripted behavior, making these institutions persistent [23, 24]. Yet, behavior can evolve over time as a result of changing regulations and norms (e.g., solving an emergent tactical purpose or when solving wicked problems). The process of standardizing procedures among members of a population from these pillars is referred to as institutional isomorphism, which is triggered by coercive, normative, and mimetic forces—constraining the ways in which individuals perform their activities [25]. This institutional isomorphism constrains the ways in which individuals perform their daily activities and cultivates expectations regarding the style of knowledge representations and production [25]. The concept of institutional isomorphism in organizational behavior theory leads to our first proposition:

Proposition 1 (homogeneity): *Data collected using unstructured-data-entry formats become homogenous within organizational units. This homogeneity is more prominent within the same organizational unit.*

The effectiveness on their decision-making is tied to the information at hand to solve such tactical purpose. This *data* homogeneity would suggest the potential for organizations to adopt standard practices in how they collect and use the information to solve a tactical need. Institutional features of organizational environments, however, can shape the actions actors take (e.g., the level of detail –specificity or focus– at which they input the information into the IS). Moreover, because of institutionalization of practice, notes from one organizational unit are similar to one another and less similar than notes from different organizational units. More importantly for the organization, is to find a way to assess the effectiveness of these unstructured notes in solving a task.

Free-text data collection's flexibility implies that the level of detail of case notes can vary across individuals across organizational units. We turn to theories from psychology to discuss the tradeoff between generalization and specification in data collection. According to psychology, categories support vital functions of an organism via *cognitive economy* and *inductive inference* [26–30]. Cognitive economy is achieved by maximally abstracting from individual differences among objects and then grouping objects in categories of larger scope [28, 31, 32]. These categories improve the ability of a person to accurately predict features of instances of a category. The trade-off between these competing functions is considered one of the defining mechanisms of human cognition and behavior [27, 33]. According to cognitive theories and theories of classification, categories (which can be represented as a class in the IS) provide cognitive economy and inferential utility, enabling humans to efficiently store and retrieve information about phenomena of interest [27, 30]. In a free-text interface these categories are not fixed and are chosen by the individual entering the data into the system.

Lukyanenko, Parsons and Wiersma [9] suggests that in a free-form data entry task, non-experts will classify more accurately at a general level than at a more specific level. When we collect structured data the level of specificity is fixed at the time of system design. Users entering unstructured data, on the other hand, can adjust to their level of specificity—by being more or less detailed [34]. Since specificity results from expertise, unstructured data collection can capture expertise better, which in turn may lead to better performance by having relevant information to support decision-making (e.g., repurposing existing data). We suggest that organizations can foster effective unstructured-data-entry practices that could result in richer data collection. We do so through the following propositions:

Proposition 2 (Inferential utility and repurposing): *Unstructured-data-entry formats can help shape effective data-entry practices in solving well-defined needs.*

Proposition 2a: Higher levels of specificity in the unstructured data collected leads to increased inferential utility.

Proposition 2b: Higher levels of specificity in the unstructured data collected facilitate repurposing data for other tasks.

The goal of the proposed design propositions is to understand the subtleties of unstructured-data-entry electronic documentation to design more effective information systems [35, 36]. The propositions enable designers to reflect on the effect of institutional practices in user generated electronic documentation. In the following sections we evaluate the propositions and provide a discussion, conclusions, and areas for future research.

4   Evaluation of Propositions

To evaluate Proposition 1 we use Stylometry, a particular application of text mining. To evaluate Proposition 2, we use text-mining techniques to analyze differences in the text authored by different case workers.

Proposition 1: Homogeneity of Data

Some researchers have argued that an author's style is comprised of a limited number of distinctive features inherent to the author, neglecting the content/context-dependency of the writing [37]. Stylometric analysis is an application of text mining that uncovers metadata from the documents and allows for statistical comparisons of these metadata as a proxy for "style". We use SAS Text Miner 9.4 to predict, based on the text in the case note, to which FCMA a particular case note belongs.

Our training set consists of all the case notes from 795 children from three agencies assigned to a mutually exclusive train and test set. We train a classification model that has the case note text and our target variable—the FCMA from which that note is coming from (e.g., FCMA A – 336 children in total, FCMA B – 213 children in total, and FCMA C – 246 children in total).

We create individual models for each FCMA and we evaluate the performance of the predictive models using commonly accepted metrics: recall, precision, and F-measure (see Table 1). Our results show that despite organizations having established guidelines of reporting, employees adopt new guidelines that become norms over time. This is reflected in how different organizational units are consistent in the way they encode home-visit notes. We also introduce the idea of organizational stylometry. To our knowledge, the use of stylometry at the population level (where many contributors to a body of text) has yet to be explored.

The results of the analysis show that by analyzing a particular case note we can predict, with a high degree of certainty, the authoring FCMA of that case note (see Table 2). These results show that each organization has its unique style, which is consistently used by its caseworkers. Based on these results we can confirm Proposition 1 that *institutional factors establish data entry practices that result in data that is similar within organizational units.*

Table 1. Evaluation metrics

Precision (P)	Recall (R)	F-measure
$P = \dfrac{TP}{TP + FP}$	$R = \dfrac{TP}{TP + FN}$	$F = \dfrac{2(P * R)}{P + R}$

Table 2. Evaluation metrics across agencies

Agency	Precision (%)	Recall (%)	F-measure (%)
Agency A	76.99	75.65	76.31
Agency B	79.81	76.15	77.94
Agency C	76.74	81.15	78.88

Proposition 2: Inferential Utility and Repurposing

To evaluate Proposition 2, we use an inductive (classification) text mining technique. First, an expert case manager provides a gold standard with labeled instances. Case notes are labeled "Yes" (uses psychotropic medication) or "No" (no use of psychotropic medication), depending on whether the child is taking psychotropic medication or not. We create individual models for each FCMA (A, B, and C) and we evaluate each within its own organizational unit (intra) and across organizational units (inter) (see Fig. 1). For each organizational unit, we assign a random sample into a training set containing 70% of the cases and a test set containing the remaining 30% of the data [15, 18]. We use SAS Text Miner 9.4 to evaluate the performance of each of the models and all the permutation comparisons across organizational units.

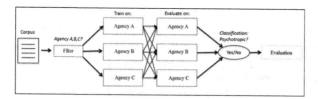

Fig. 1. Intra and Inter-agency data mining process

There is no standard definition of what a substantial difference in F-measure improvement should be. In the field of information retrieval a 5% performance improvement is considered a substantial improvement [38, 39]. The z-test for proportions evaluates the statistical difference between two population proportions p_1 and p_2 [40, 41]. To test the difference between proportions we compute the following:

$$z_{proportions} = \frac{\bar{p}_1 - \bar{p}_2}{\sqrt{\bar{p}(1 - \bar{p})\left(\frac{1}{n_1} + \frac{1}{n_2}\right)}} \tag{1}$$

We evaluated each FCMA by comparing the performance when tested with data from the same organizational unit (intra-FCMA) and compared to models that use data from other organizational units (inter-FCMA). We highlight in bold any statistically significant differences for precision and recall using a z-test for proportions (two-tailed test at the 95% confidence level). We consider the difference in F-measure as substantial if the difference between F-measures is more than 0.05 and the difference in precision or recall is statistically significant (determined using the z-test for proportions and highlighted in bold and with a * symbol) [38].

Table 3 shows that the differences in F-measure are substantial in five out of the six pairs. The results show that two of the agencies (FCMA A and FCMA C) consistently perform better in classifying cases of psychotropic drug use. This shows that unstructured data entry formats may result in differences in how information is collected across different organizational units in the organization. Institutional theory helps explain how institutional factors shape practices by individuals in different organizational units, and how these practices can become stable over time and adopted by other individuals, making practices persistent. This validates our first proposition that *data collected using unstructured-data-entry formats become homogeneous within organizational units.*

Common NLP tools include document tokenizing, stemming, parts-of-speech tagging, noun group extraction, applying stop lists, entity identification, and multiword terms handling [42]. The document is parsed and tagged based on the syntactical relationship between terms –based on the position in a sentence and rules of grammar [43].

Table 3. Difference between proportions for precision (P) and Recall (R)

Train	Evaluation	Precision	Recall	F-Measure
FCMA A	FCMA A	78.57	70.97	74.58
	FCMA B	65	52.7	58.21
	FCMA C	31.94	30.67	31.29*
	Z-Value (FCMA A-FCMA B)	1.2858	1.7303	
	Z-Value (FCMA A-FCMA C)	4.2082 (p<0.01)	3.9911 (p<0.01)	
FCMA B	FCMA B	46.15	54.54	50
	FCMA A	45.59	30.69	36.69*
	FCMA C	32	21.33	25.6*
	Z-Value (FCMA B-FCMA A)	0.1377	2.1261 (p<0.05)	
	Z-Value (FCMA B-FCMA C)	1.1864	3.0229 (p<0.01)	
FCMA C	FCMA C	64.71	50	56.41
	FCMA A	33.33	16.83	22.37*
	FCMA B	59.26	21.62	31.68*
	Z-Value (FCMA C-FCMA A)	2.2762 (p<0.05)	3.3621 (p<0.01)	
	Z-Value (FCMA C-FCMA B)	0.3613	2.5992 (p<0.01)	

The aim is to convert human language into formal representations computers can manipulate, including part-of-speech tagging (POS), POS sequences, or n-gram models [42, 44, 45]. Because authors do not always follow grammatical rules, the complexity of multiple meanings for words, and the domain specific use of vocabulary may require some additional considerations.

Results show that there is no statistical significant difference between the full model and the model that has no Part-of-Speech and Noun Group features but does have a term weighting scheme (Term Frequency, Term Weight). Consistent with previous research, the terms used are a more salient factor of prediction compared to the language structure of a case note. Human language is subtle, with many unquantifiable yet salient qualities. Users with different levels of expertise tend to produce information that differs in quality and level of abstraction. For example, within the category "taking medication", a conceptual hierarchy can be the following: (a) medication (b) psychotropic medication (c) Lisdexamfetamine (d) Vyvanse, which goes from the most general (a) to the most specific (d). Knowing a child is taking Vyvanse (d) gives more information than just knowing a child is taking medication (a).

We assess language use (in terms of structure and meaning of the case notes) by including/excluding natural language processing (NLP) features.

A case note authored by Agency A "Takes mg of vyvanse by mouth once a day [...] she has to call the doctor to schedule for a refill" has a confidence of 0.953 of being psychotropic drug use case. Whereas the following case note authored by Agency C "child has an immune system medical condition that requires many medications to keep her healthy" has a confidence of 0.594 of being a case of psychotropic drug use –as measured by the singular vector decomposition scores. The results of the analysis show that higher levels of specificity in the data collected leads to increased inferential utility, which can ultimately help the organization solve unanticipated tasks using these data. This validates Proposition 2 that *higher levels of specificity in the unstructured data collected leads to increased inferential utility, which can in turn be leveraged to repurpose data for a different task it was originally designed for.*

In the final section we discuss the implications of our findings for theory and practice.

5   Implications for Research and Practice

Our findings have important implications both to theory and practice of conceptual modeling, unstructured information collection, text mining and business analytics, and general organizational data management.

We believe our work is timely. Traditional information systems research has been concerned with finding similar elements in highly structured data sets [46] and the study of unstructured data sources is a relatively recent active stream of work. At the same time, unstructured data sources continue to grow in prominence fueled by the explosive growth in social media and online content production which tends to be text-based. Our work aims to provide both theoretical and practical insight into the nature of unstructured information. The arguments and findings of our work are thus applicable to user generated content settings and as we as our context of corporate unstructured data [13]. Indeed, researchers continue to call for novel approaches to structure user generated content to make it more consistent and usable in organizational analysis [11]. Our work has strong potential to contribute to the efforts to make user generated content more usable by increasing its potential for reuse. In the future, we hope to extend our work to the area of user generated content (specifically, crowd-sourcing) to address the issue of repurposing it for unanticipated insights.

In our paper, we show that we can reliably detect organizational styles. This insight can be used to improve organizational processes and foster more effective data reuse. First, our research suggests that the data-entry formats of the information system can highlight the existence of different organizational styles across organizational units. Second, our research suggests that the flexibility of free-form data entry motivates individuals to stay truthful to their organizational unit's reporting expectations. This highlights the trade-off between different data-entry formats and the data collected by the organization.

Our results demonstrate the role of the level of specificity in enabling unanticipated insights. The results of this study can be generalized to other domains and can provide insight to effective system design—the effect of particular designs (that are more/less flexible). In a fully structured scenario, the user is guided by the interface on what needs to be reported. In a semi-structured scenario, pre-established templates guide data entry but allows for some deviation by the user to input something not related to a particular template. In an unstructured scenario (e.g., free-form), the individual has the liberty to enter data, which is typically defined by the organization (e.g., business processes, training).

Our research encourages experts to be as specific as they can while allowing non-experts to input information at a more general level. Higher specificity, however, requires higher expertise. Thus, it may hinder collaboration from non-experts. Future work should focus on how these different data-entry formats may preclude the collection of valuable information (leading to information loss) when both novices and experts contribute to the system. Previous research have shown that limiting data-entry to experts can preclude the input of valuable information from non-experts and can lead to data accuracy problems [9].

Our results are consistent with psychology research in that the level of specificity of the information limits the applications for which that data can be used. In an environment where individuals have a similar level of expertise based on their background and training, it is preferable for them to be more specific when they enter the data into the system (e.g., the child is taking 5 mg of Adderall provides more information than just saying the child is taking medication). A practical implication to this is that depending on whether the individuals looking at the text are a non-experts vs. experts, the individual writing the text can choose to contribute beyond what he believes is the information required for the reader. This allows for increased inferential utility that can prove beneficial when dealing with unanticipated uses of the data.

Our study also provides guidance of the implications of choosing how data entry formats of a system are designed—and what is it that they would like to capture from their users. To the best of our knowledge, Authorship Analysis had only been done at the individual level. We extend this analysis for authorship identification at the group level (e.g., identifying the authoring organizational unit of a body of text). This can be used by organizations to assess the consistency of data-entry practices in an organization and can be extended by using analytical techniques to create dimensions of categories these documents fall into or metrics that relate to reliability of the data.

The tension between data collection at different levels of granularity further suggests exciting new opportunities at the intersection of conceptual modeling and data analytics. Conceptual modeling research has long studied the nature of content aggregation, part – whole relationships and the general ontological assumptions behind data collection [47–49]. These can become valuable sources of guidance for innovative analytics approaches aiming at drawing inferences from data collected at different levels of analysis. We hope to pursue this work in the future.

6 Limitations

This study is not without limitations. There is a threshold for the classification models accuracy that is directly related to the quality of the data in the gold standard. For instance, psychotropic medication was attributed to the foster home and not the child. If a foster home has multiple children and one was taking psychotropic medication, all of these children would appear as taking psychotropic medication and vice-versa. This is a limitation that introduces biases in the classification models. Moreover, we did not take into account time windows (e.g., a kid that was prescribed psychotropic medication is no longer taking that medication). However, this does not undermine the goal of our work, which is to understand the relationship of data-entry practices in repurposing data. Future work should focus in providing a method to evaluate when using data in the aggregate is justified as opposed to highlighting meaningful segments for separate analysis.

References

1. Gantz, J., Reinsel, D.: Extracting value from chaos. IDC Iview **1142**, 1–12 (2011)
2. Boudreau, M.-C., Robey, D.: Enacting integrated information technology: a human agency perspective. Organ. Sci. **16**, 3–18 (2005)
3. Wand, Y., Weber, R.: On the deep structure of information systems. Inf. Syst. J. **5**, 203–223 (1995)
4. DeSanctis, G., Poole, M.S.: Capturing the complexity in advanced technology use: adaptive structuration theory. Organ. Sci. **5**, 121–147 (1994)
5. Burton-Jones, A., Grange, C.: From use to effective use: a representation theory perspective. Inf. Syst. Res. **24**, 632–658 (2012)
6. Berg, M., Goorman, E.: The contextual nature of medical information. Int. J. Med. Inform. **56**, 51–60 (1999)
7. Berg, M.: Implementing information systems in health care organizations: myths and challenges. Int. J. Med. Inform. **64**, 143–156 (2001)
8. Eveleigh, A., Jennett, C., Blandford, A., Brohan, P., Cox, A.L.: Designing for dabblers and deterring drop-outs in citizen science. In: Proceedings of the 32nd Annual ACM Conference on Human Factors in Computing Systems, pp. 2985–2994. ACM (2014)

9. Lukyanenko, R., Parsons, J., Wiersma, Y.F.: The IQ of the crowd: understanding and improving information quality in structured user-generated content. Inf. Syst. Res. **25**, 669–689 (2014)
10. Van Kleek, M.G., Styke, W., Karger, D.: Finders/keepers: a longitudinal study of people managing information scraps in a micro-note tool. In: Proceedings of the SIGCHI Conference on Human Factors in Computing Systems, pp. 2907–2916. ACM (2011)
11. Lukyanenko, R., Parsons, J., Wiersma, Y., Wachinger, G., Huber, B., Meldt, R.: Representing crowd knowledge: guidelines for conceptual modeling of user-generated content. J. Assoc. Inf. Syst. **18**, 2 (2017)
12. Jabbari Sabegh, M.A., Lukyanenko, R., Recker, J.C., Samuel, B., Castellanos, A.: Conceptual modeling research in information systems: what we now know and what we still do not know (2017)
13. Burton-Jones, A., Volkoff, O.: How can we develop contextualized theories of effective use? A demonstration in the context of community-care electronic health records. Inf. Syst. Res. (2017)
14. Lukyanenko, R., Parsons, J.: Information quality research challenge: adapting information quality principles to user-generated content. J. Data Inf. Qual. (JDIQ) **6**, 3 (2015)
15. Tremblay, M.C., Berndt, D.J., Luther, S.L., Foulis, P.R., French, D.D.: Identifying fall-related injuries: text mining the electronic medical record. Inf. Technol. Manage. **10**, 253–265 (2009)
16. Sørlie, T., Perou, C.M., Tibshirani, R., Aas, T., Geisler, S., Johnsen, H., Hastie, T., Eisen, M.B., van de Rijn, M., Jeffrey, S.S.: Gene expression patterns of breast carcinomas distinguish tumor subclasses with clinical implications. Proc. Natl. Acad. Sci. **98**, 10869–10874 (2001)
17. Larsen, K., Bong, C.H.: A tool for addressing construct identity in literature reviews and metaanalyses. MIS Q. **40**, 529–551 (2016)
18. Castillo, A., Castellanos, A., Tremblay, M.C.: Improving case management via statistical text mining in a foster care organization. In: Tremblay, M.C., VanderMeer, D., Rothenberger, M., Gupta, A., Yoon, V. (eds.) DESRIST 2014. LNCS, vol. 8463, pp. 312–320. Springer, Cham (2014). doi:10.1007/978-3-319-06701-8_21
19. Luther, S., Berndt, D., Finch, D., Richardson, M., Hickling, E., Hickam, D.: Using statistical text mining to supplement the development of an ontology. J. Biomed. Inform. **44**, S86–S93 (2011)
20. Jepperson, R.L.: Institutions, institutional effects, and institutionalism. New Institutionalism Organ. Anal. **6**, 143–163 (1991)
21. Giddens, A.: Central Problems in Social Theory: Action, Structure, and Contradiction in Social Analysis. University of California Press, Berkeley (1979)
22. Sewell Jr., W.H.: A theory of structure: Duality, agency, and transformation. Am. J. Soc. **98**, 1–29 (1992)
23. Hughes, E.C.: The ecological aspect of institutions. Am. Sociol. Rev. **1**, 180–189 (1936)
24. Barley, S.R., Tolbert, P.S.: Institutionalization and structuration: Studying the links between action and institution. Organ. Stud. **18**, 93–117 (1997)
25. DiMaggio, P.J., Powell, W.W.: The iron cage revisited: institutional isomorphism and collective rationality in organizational fields. Am. Soc. Rev. **48**(2), 147–160 (1983)
26. Lakoff, G.: Women, Fire, and Dangerous Things. University of Chicago Press, Chicago (1987)
27. Roach, E., Lloyd, B.B., Wiles, J., Rosch, E.: Principles of categorization (1978)
28. Smith, E.E., Medin, D.L.: Categories and Concepts. Harvard University Press, Cambridge (1981)
29. Smith, E.E.: Concepts and thought. In: The Psychology of Human Thought, p. 19 (1988)

30. Parsons, J.: An information model based on classification theory. Manage. Sci. **42**, 1437–1453 (1996)
31. Fodor, J.A.: Concepts: Where Cognitive Science Went Wrong. Clarendon Press, Oxford (1998)
32. Murphy, G.L.: The Big Book of Concepts. MIT Press, Cambridge (2004)
33. Corter, J., Gluck, M.: Explaining basic categories: feature predictability and information. Psychol. Bull. **111**, 291–303 (1992)
34. Lukyanenko, R., Castellanos, A.: Introducing information gradient theory. In: Breakthroughs and Emerging Insights from Ongoing Design Science Projects: Research-in-progress papers and poster presentations from the 11th International Conference on Design Science Research in Information Systems and Technology (DESRIST 2016) 2016, St. John, Canada, 23–25 May (2016)
35. Walls, J.G., Widmeyer, G.R., El Sawy, O.A.: Building an information system design theory for vigilant EIS. Inf. Syst. Res. **3**, 36–59 (1992)
36. Eisenhardt, K.M.: Building theories from case study research. Acad. Manag. Rev. **14**, 532–550 (1989)
37. De Vel, O., Anderson, A., Corney, M., Mohay, G.: Mining e-mail content for author identification forensics. ACM Sigmod Rec. **30**, 55–64 (2001)
38. Adomavicius, G., Sankaranarayanan, R., Sen, S., Tuzhilin, A.: Incorporating contextual information in recommender systems using a multidimensional approach. ACM Trans. Inf. Syst. (TOIS) **23**, 103–145 (2005)
39. Sparck Jones, K.: Automatic indexing. J. Doc. **30**, 393–432 (1974)
40. Kachigan, S.K.: Statistical Analysis: An Interdisciplinary Introduction to Univariate & Multivariate Methods. Radius Press, New York (1986)
41. Fleiss, J.L., Levin, B., Paik, M.C.: Statistical Methods for Rates and Proportions. Wiley, New York (2013)
42. Manning, C.D., Schutze, H.: Foundations of Statistical Natural Language Processing. MIT Press, Cambridge (1999)
43. Berry, M.W., Castellanos, M.: Survey of text mining. Comput. Rev. **45**, 548 (2004)
44. Abbasi, A., Chen, H.: CyberGate: a design framework and system for text analysis of computer-mediated communication. Mis Q. **32**(4), 811–837 (2008)
45. Holmes, D.I.: The evolution of stylometry in humanities scholarship. Literary Linguist. Comput. **13**, 111–117 (1998)
46. Batini, C., Lenzerini, M., Navathe, S.B.: A comparative analysis of methodologies for database schema integration. ACM Comput. Surv. (CSUR) **18**, 323–364 (1986)
47. Shanks, G., Tansley, E., Nuredini, J., Tobin, D., Weber, R.: Representing part-whole relationships in conceptual modeling: an empirical evaluation (2002)
48. Evermann, J., Wand, Y.: Towards ontologically based semantics for UML constructs. In: Kunii, H.S., Jajodia, S., Sølvberg, A. (eds.) ER 2001. LNCS, vol. 2224, pp. 354–367. Springer, Heidelberg (2001). doi:10.1007/3-540-45581-7_27
49. Wand, Y., Storey, V.C., Weber, R.: An ontological analysis of the relationship construct in conceptual modeling. ACM Trans. Database Syst. (TODS) **24**, 494–528 (1999)

Analysis of Benefits for Knowledge Workers Expected from Knowledge-Graph-Based Information Systems

Mariia Rizun[1]([✉]) [iD] and Vera G. Meister[2] [iD]

[1] University of Economics in Katowice, Katowice, Poland
mariia.rizun@ue.katowice.pl
[2] Brandenburg University of Applied Sciences,
Brandenburg an der Havel, Germany
vera.meister@th-brandenburg.de

Abstract. The paper raises the issues of Knowledge Workers job and the Knowledge Management Information Systems (KMIS) that support it. The first objective of the paper is to define the notions Knowledge Worker and KMIS - in the form of concept maps. In the paper, the knowledge actions conducted by the Knowledge Workers are also analyzed and the most frequently performed actions are distinguished. The two actions - analysis and information organization, are used as the basis for examination of KMIS quality in terms of Knowledge Workers support. A case study of Moodle platform as a KMIS supporting actions of Knowledge Workers in higher education justifies the conclusions of the support quality of traditional KMIS. With the objective to suggest improvement of the KMIS the authors introduce a design approach for Knowledge-Graph-based Information Systems (KGIS) and describe the general benefits of them, again referring to such knowledge actions as *analysis* and *information organization*. The paper also covers the limitations that might relate to a broad implementation of KGIS, connected with infrastructure, technology, organization, and media effects. At the end, the authors present another small case study dealing with the expected support quality of a semantic IT service catalog, prototypically implemented in a German higher education institution.

Keywords: Knowledge management · Knowledge worker · Knowledge management information system · Knowledge-graph-based information system

1 Introduction

In addition to globalization, hyper competition and technological development, organizations all over the world are increasingly facing the challenges of digitalization. This kind of environment requires a different way of thinking and behavior for those who want to stay competitive and prosperous. That is why most of the organizations are constantly seeking for new sources of advantage. In the last 20+ years, the factors influencing organizational competitiveness have transformed from physical and tangible resources to those based on knowledge. Thus, organizations become more and

© Springer International Publishing AG 2017
S. Wrycza and J. Maślankowski (Eds.): SIGSAND/PLAIS 2017, LNBIP 300, pp. 25–39, 2017.
DOI: 10.1007/978-3-319-66996-0_3

more dependent on the successful definition, application and integration of their knowledge management (KM) processes [1, 2].

It is generally accepted that since knowledge is a critical factor for an organization's survival, it should be captured, managed and utilized in a way that fosters organizational development. Knowledge management processes are viewed as encompassing all activities that create or locate knowledge, manage the flow of knowledge and ensure that knowledge is used effectively and efficiently for the long-term benefit of the organization. Therefore, effective organization of knowledge is in the heart of KM and the significance of KM lies in the most gainful use of knowledge for organizational purposes [3].

The group of people engaged in knowledge management processes are called the knowledge workers (KWs). There are three key features, which differentiate knowledge work from other forms of conventional work. Firstly, the KW adds value to work through mental activities. Secondly, the kind of thinking applied by KWs is not a step-by-step linear mental work – they must be creative and non-linear in their thinking. The third distinctive feature of knowledge work is that it uses knowledge to produce more knowledge [4].

Implementation of various IT systems for supporting knowledge management processes has been of significant popularity in the early years of our century. Knowledge Management Information Systems (KMIS) have been supposed to integrate functionalities and provide comprehensive organization and control of the processes of knowledge generation, transfer and storage [5]. Information technologies were expected not only to form the ground for KWs to apply and process knowledge but even to replace them to some extent by implementing semantic technologies. A lot of effort was dedicated to research and development of so-called expert systems.

Nevertheless, the most popular information technology used by KWs up to now are standard Office tools, like e-mail, text processing, spreadsheet calculation and presentation editing. That is why, when studying the activity of knowledge workers, it is reasonable to pay attention to the tools, which shall assist them. It is necessary to examine the functionalities of these KMISs and analyze, whether they fully meet the requirements of KWs and what improvements can be made to these systems' design and operation. The paper makes a timely contribution to this examination. It primarily addresses decision makers in enterprises and organizations concerned with management, structure and technological tool support for KWs work environment as well as with forming according competencies within their staff.

2 Research Objectives and Methodology

From the above-stated there come the following objectives of the paper:

1. With the purpose to clarify the subjects of the research and to set the specific area of further research – to define the notions *Knowledge Worker* and *Knowledge Management Information System*.
2. With the objective to specify the demands of KWs as for the software tools that assist them in the process of knowledge creation, storage, distribution etc. – to determine the actions performed by them; specify a *classification of KW actions*.

3. To clarify the relevance of semantic technologies for KMIS – to analyze the support quality of traditional KMIS for selected, important KW actions; to introduce a design approach for *Knowledge-Graph-based Information Systems* (KGIS) and to derive the *benefits in KW action support* provided by these systems.

Realization of the objectives is structured as follows: Sect. 2 firstly gives a deeper definition of the notion Knowledge Worker. Considering the research on roles and typical actions of KWs and analyzing the core processes of Knowledge Management, the classification of KWs actions is elaborated. Section 3 is dedicated to the analysis of the support quality provided by traditional KMIS systems with respect to selected action types of KWs elicited before. The results of this general analysis are illustrated and confirmed by a case study performed in the field of higher education. Section 4 focusses on Knowledge-Graph-based Information Systems. First a definition of knowledge graph is specified from literature analysis. Due to this operational definition, a comprehensive design approach for KGISs is suggested. Rationales for expecting better support quality for selected KW actions by KGISs are given. Again, a case study is referred to confirm these results. Finally, the impediments for the implementation and maintenance of KGIS are discussed. In the last section conclusions on the research are drawn and some further steps of the authors' research are described. All sections of the paper are based on proper literature analysis, some reuse findings from other works of the authors.

3   Knowledge Workers and Their Activities

Presently there exist many versions of definition of a knowledge worker. Some of them are similar and express the same ideas in other words, and some of them present a different vision on the KW phenomena. To clarify the subject of the research the authors find it reasonable to examine the existing definitions and to formulate an aggregated definition.

The most valuable (from the authors' point of view) *definitions of a knowledge worker* are presented as follows:

- KWs are high level employees who apply theoretical and analytical knowledge, acquired through formal education, to develop new products or services [6].
- A KW performs a set of knowledge-intensive tasks (decision-making, knowledge-production scenarios, monitoring organizational performance, etc.), mainly with the support of IT, dominated by communication, data production and consumption actions [7].
- KWs understand, define, influence and help shape their domain of influence, knowledge, activity and responsibility. They understand the people, information and potential resources within that domain as well as have the authority to act within that domain [8].
- For KWs the role of knowledge is central to the job, they must be highly educated or expert. KWs like autonomy, they usually have good reasons for doing what they do, they value their knowledge and don't share it easily (comp. [9]).

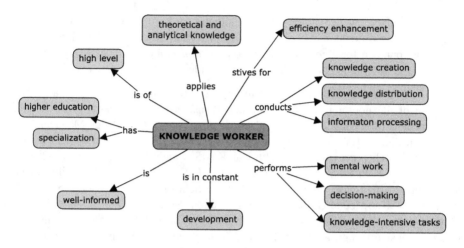

Fig. 1. Definition of knowledge worker as concept map

Based on the definitions given above, as well as those not mentioned in the paper (including [4, 10–12]) the authors have formed an aggregated definition of a knowledge worker (comp. [13]) and have chosen to present it here as a concept map (Fig. 1). The map includes key words from the analyzed definitions, presenting the key characteristics of a knowledge worker.

Table 1. Knowledge workers roles and actions [7]

Role	Description	Typical actions
Controller	monitors the organizational performance based on raw data	analysis, dissemination, information organization, monitoring
Helper	transfers information to teach others, once he or she passed a problem	authoring, analysis, dissemination, feedback, information search, learning, networking
Sharer	disseminates information in a community	authoring & co-authoring, dissemination, networking
Learner	uses information and practices to improve personal skills and competencies	acquisition, analysis, expert search, information search, learning, service search
Linker	associates and mashes up information from different sources to generate new information	analysis, dissemination, information search, information organization, networking
Networker	creates personal or project related connections with people involved in the same kind of work, to share information and support each other	analysis, dissemination, expert search, monitoring, networking, service search
Organizer	is involved in personal or organizational planning of activities, e.g. to-do lists and scheduling	analysis, information organization, monitoring, networking
Retriever	searches and collects information on a given topic	acquisition, analysis, expert search, information search, information organization, monitoring
Solver	finds or provides a way to deal with a problem	acquisition, analysis, dissemination, information search, learning, service search
Tracker	monitors and reacts on personal and organizational actions that may become problems	analysis, information search, monitoring, networking

Thus, it is reasonable to say that the first part of *the first objective of the paper* – definition of a knowledge worker, has been reached.

For realization of *the second objective of the paper* – specification of the classification of KWs actions, it is necessary to refer to the work of Reinhardt et al. [7]. The work contains the description of ten *knowledge worker roles*, each mapped to a specific subset of overall twelve different *knowledge worker actions* (Table 1).

The focus on KWs' roles indicates as most frequently listed actions the following: analysis, dissemination, information search, and networking (for details see the numbers in parenthesis in Fig. 2).

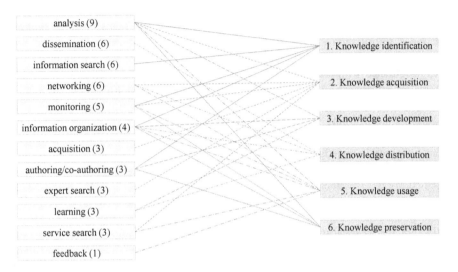

Fig. 2. Specified classification of KWs' actions related to core processes of KM

Coming back to KM processes as field of action for KWs a reference framework provided by [14] is chosen. It encompasses six core processes of KM with corresponding leading questions:

1. *Knowledge identification*: How to create internal and external transparency about knowledge existing in the organization?
2. *Knowledge acquisition*: What skills the organization shall acquire from outside?
3. *Knowledge development*: How to create new knowledge for the organization?
4. *Knowledge distribution*: How to bring the knowledge to the right place?
5. *Knowledge usage*: How to ensure the knowledge application organization-wide?
6. *Knowledge preservation*: How to protect the organization from loss of knowledge?

Analyzing the descriptions given in [14], the authors propose a specified classification of KWs actions with respect to core processes of KM in Fig. 2. At this stage, it is necessary to remember about the second part of *the first objective of the paper* – the definition of a Knowledge Management Information System, which follows in the next section.

4   Knowledge Management Information Systems

In [2] there is an opinion that technical systems (for instance, groupware, e-mail, databases, intranet etc.) are applied in KM for collection, codification, storage, and manipulation of knowledge.

In [4] it is stated that technologies are used by KWs to facilitate access to information and its manipulation. Information technologies (IT) are referred to as computer equipment and programs applied to access, process, store, and disseminate information. As examples word processing, spreadsheet, and electronic mail programs are mentioned. The paper also highlights the fact that IT are designed to reduce the amount of time workers spend on information access, management and manipulation and to increase the accuracy of these processes and, at the same time, that IT and the Internet have made information easy-to-access, user-friendly and up-to-date. Moreover, the emerging mobile and wireless IT supports the mobile nature of the KWs jobs.

In accordance with the work of Maier and Hädrich [15] Knowledge Management Systems are characterized by such terms as: knowledge warehouse, knowledge management software, technology or organizational memory (information) system, e-learning suite, learning management platform, portal, document management, collaboration or groupware. A KMIS is defined as a system that functions to provide a means by which knowledge from the past is brought to bear on present activities, thus resulting in increased levels of effectiveness for the organization. The concept map presented in Fig. 3 visualizes general requirements and features for KIMS based on the definitions above as well as on the authors' experience.

Since there is a big and ever-changing variety in systemizing and designating KMIS, not the systems itself with their ambiguous names shall be investigated, but instead the required functionalities of the systems with respect to the KW actions under consideration on the one hand and the information technology paradigm applied in a

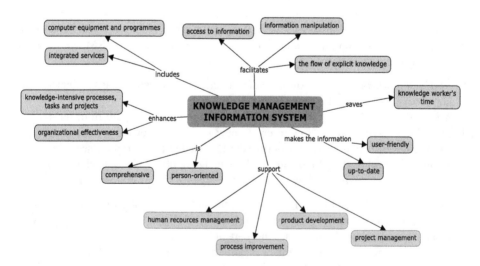

Fig. 3. KMIS general requirements and features

KMIS on the other hand. According to the classification given in Fig. 2, the two KW actions *analysis* and *knowledge organization* can be evaluated as crucial, since they relate to almost all the core processes of KM. It should be recalled that analysis was also the most frequent action in the context of KW roles. It therefore seems reasonable to limit the following investigations to these two actions.

In the following the notions of data, information and knowledge are used like defined by Davenport and Prusak in [1]. Shortly spoken, they stated (i) data as discrete, objective facts about events, (ii) information as messages, usually in the form of a document, and (iii) knowledge as a fluid mix of framed experience, values, contextual information, and expert insight originating from the minds of knowers and often becoming embedded in documents or repositories but also in organizational routines, processes, practices, and norms.

KW action – analysis. Analysis is a holistic, systematic investigation, in which the object under investigation is dissected and divided into its components, and these are then ordered, examined and evaluated, considering also the relationships of the individual elements and their integration [16]. When transferring this general definition to the domain of knowledge management, the following elementary actions can be identified: (i) drill down, filter, order and compare data; (ii) manipulate or aggregate data; (iii) visualize the structure and relations between data, information and/or knowledge.

KW action - information organization. This KW action pursue the goal to provide the knowledge of an organization in a meaningful, maintainable and easy to access as well as secure way. This requires at least some of the following elementary actions: (i) develop, specify and maintain a conceptualization together with a categorization or classification; (ii) arrange data, information and/or knowledge according to the implemented conceptualization; (iii) provide additional tags, links or formal relations to some other piece of data, information or knowledge; (iv) document knowledge sources, responsibilities and maintenance processes; (v) implement access rights to data, information and/or knowledge.

Information technology paradigm. Data, information and knowledge can also be viewed as a classification of the information technology paradigm applied in a KMIS. Even if a strict differentiation is problematic in practice, a predominant paradigm can be identified for most of the systems. In this section, the investigation of the support quality for the two selected KW actions will be restricted to KMIS following the data or the information paradigms which are currently predominating, whereas the knowledge paradigm is subject of Sect. 4.

A KMIS shall be considered as applying a data paradigm when it is basically founded on a traditional (relational) database. Since most of the analysis actions are data-oriented it enables a valuable support of these actions given that (i) the database schema is sufficiently rich and (ii) there are enough features provided for customization and/or visualization of analysis results. Nevertheless, KMIS of this category usually have little flexibility regarding changed structures or conditions in organizations. They can implement an appropriate information structure at a dedicated moment in time (mostly at the date of acquisition), but it is costly or cumbersome or even impossible to use them in an agile information organization process. Examples of data-centered systems are Enterprise Resource Planning or Customer Relationship Management systems as well as Business Intelligence Systems and Data Warehouses.

Typical information-centered systems are Document Management Systems, Wikis or Enterprise Content Platforms, like e.g. e-learning Platforms for education institutions. Main entities in this kind of systems are documents in the sense of information elements which can be arranged (i) in an object schema, mostly predefined, at least partly, and (ii) in a hierarchical conceptual structure, or at least in lists. The predefined object schema is often restricted to general object types, like users, teams or groups; projects, courses or programs; and subject-related entity types, like issues, tasks, or learning activities. Information organization in such systems is supported on the on hand in the frame of the object schema, but more valuable by the conceptual hierarchy which can be enlarged, deepened, rearranged etc. Some systems support the inter-linking of documents or the adding of tags. Nevertheless, the links are mostly not specifiable and shall be interpreted as see-also relations, whereas the tags may lack of object identity because of the usage of different spelling or forms of words, as well as undeclared synonyms. Regarding the KW action analysis, these systems provide mostly a poor support quality. Most of the information is stored in unstructured form, main recipients of information are expected to be humans. Some statistics of metadata according to the object schema may be provided.

Table 2 summarizes the results of the discussion on the support quality of the system paradigms regarding the KW actions analysis and knowledge organization.

Table 2. Support quality of KMIS regarding selected KW actions

KMIS paradigms and examples	Support of the KW action analysis	Support of the KW action information organization
Data paradigm - ERP systems - Business intelligence - Data warehouses	Good or very good	Weak, costly or cumbersome
Information paradigm - CMS - Wikis - ECM platforms	Poor or very poor	Good – depending on the use case

In order to support the conclusions on the support quality of KMISs, the authors refer to a case study examining an e-learning platform for education institutions - the open-source platform Moodle. Developed on pedagogical principles, Moodle is used for blended learning, distance education and other e-learning projects in schools, universities, workplaces and other sectors. Moodle is a data-centered KMIS to create private websites with online courses for educators and trainers. Technical features of Moodle allow to obtain, share, store and apply information, as well as having it rather secure in Moodle's storage system.

Table 3. KW requirements for changes to Moodle features

Change requirements to Moodle features	Answers
Change the structure of catalog of files	31,8%
Implement a key word search of materials	52,4%
Increase the level of materials protection	13,6%
Implement a live chat with students and other users of the platform	23,8%
Add calendar settings: reminders for users to add new material	33,3%

The questionnaire was developed with an objective to find out whether Moodle meets all the requirements of KWs [13]. It was aimed at teachers of higher education institutions and distributed in two large Polish universities. The respondents of the research possess the degree of Master of Science (30,2%), Doctor of Philosophy (60,5%) and the Habilitation degree (9,3%). It was found out that 59,1% of the respondents use the platform during the whole study process, while 4,5% stated that they use the platform on rare occasions.

Among others, the questionnaire addressed the issue of changes to be introduced to Moodle suggested by the teachers to optimize their work and attract more users. More than one issue could be chosen. Table 3 contains the results demonstrating weaknesses regarding the support quality of Moodle as a KMIS.

In detail, the necessity to change the structure of catalogues and to add a keyword search means that the data are not arranged in a proper way. The need to increase the safety level refers to the implementation of access restrictions to data or documents. Thus, it can be stated, that the level of information organization at Moodle requires some improvements. At the same time, suggestion about live-chat for the teachers and students means that the options to manipulate data and transfer information are evaluated as too weak – i.e. the level of analysis support also could be improved.

Thereby the notion of KMIS is defined and the support quality of traditional KMIS for two important KW actions is analyzed in general and illustrated by a case study in the field of higher education. Thus, the *second part of the first objective* and the *first part of the third and last objective* of the paper are achieved. For completion, it remains to meet *the rest of the third objective* – to introduce a design approach for Knowledge-Graph-based Information Systems (KGIS) and to derive the benefits in action support provided by such systems, which is subject of Sect. 4.

5 Expected Benefits from KGIS

The elaborations in Sect. 3 have shown that KMISs designed according to the data-centered or information-oriented paradigm fall short in comprehensive support of important KW actions. Up to now most of the KMIS implement a hybrid of both paradigms but with an emphasis to one of the two. Nevertheless, the vision of a really knowledge-based KMIS is not new. Research under the headlines of artificial intelligence, ontology engineering, semantic technologies and related subjects is going on for more than 30 years. An introductory summary on this topic can be found e.g. in [17].

Since knowledge according to [1] is a "fluid mix of framed experience, values, contextual information, and expert insight originating from the minds of knowers" the architecture of knowledge-based information systems must implement a basic structural artifact reflecting these characteristics: it shall be fluidly mixable and must reflect the patterns of mental operations. It is common sense or at least state of the art that graph structures are best suited to these requirements. Research on Knowledge Graphs (KG) or similar concepts started in the 80th of the last century (see e.g. [18, 19]). Whereas these theories were mostly influenced by mathematical methods, starting from 2012 the notion of KG is associated predominantly with domain-specific knowledge bases implemented by the big players in IT industry: Google, Facebook and Microsoft, to name some of them.

Within the multitude of recent publications on KGs [20–23] the definition of Paulheim e.a. [24] is chosen because of its clear structure and high viability: "A knowledge graph (i) mainly describes real world entities and their interrelations, organized in a graph, (ii) defines possible classes and relations of entities in a schema, (iii) allows for potentially interrelating arbitrary entities with each other and (iv) covers various topical domains."

Due to the mentioned characteristics of a KG and to recent developments in Web technology (particularly open standards for APIs and user interfaces) knowledge-based information systems can be designed and implemented as a fluid mix of knowledge resources, orchestration and management processes, knowledge engineering and querying facilities, data storages, and a big variety of fine-grained, specialized knowledge services. A KMIS applying such a design approach is to be called Knowledge-Graph-based Information System (KGIS). In [25] was developed an architectural design (Fig. 4) for a KGIS in the domain of higher education institutions, which can be easily transferred to other domains.

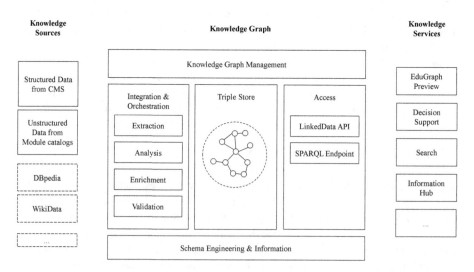

Fig. 4. EduGraph architecture representative for KGIS [25]

To assess the support quality of KGIS for the two selected KW actions, each of their elementary actions worked out in Sect. 3 will be examined in detail. The results of this examination are collected in Table 4.

Table 4. Support quality of KGIS for the KW actions *analysis* and *information organization*

KW action	Elementary actions	Rationales for KGIS support quality
Analysis	Drill down, filter, order and compare data	Multi-faceted support by SPARQL SELECT queries using in addition filtering, ordering and other functions
	Manipulate or aggregate data	Flexible support by connection to various external data sources and by SPARQL queries using aggregate functions
	Visualize the structure and relations between data, information and/or knowledge	Constantly increasing support by standardized tools and features like maps, timelines, explorable network diagrams etc.
Information organization	Develop, specify and maintain a conceptualization together with a categorization/classification	Depends on the editing tool features of the schema engineering component – ranks from thesaurus management to modeling of complex ontologies, user experience shall be improved
	Arrange data, information and/or knowledge according to the implemented conceptualization	Natively supported by basic technological elements like URI, RDF and implemented standard vocabularies
	Provide additional tags, links or formal relations to other data, information or knowledge	Like above + support of individual customization by domain-specific relations
	Document knowledge sources, responsibilities and maintenance processes	Adjustable support by specific standard vocabularies and technologies like PROV-O for data provenance, SHACL for schema constraints, BPMN for process execution and control
	Implement access rights to data, information and/or knowledge	Support by SPARQL CONSTRUCT queries based on access features as part of schema; integration with organizational identity management is recommended

To subsume the results collected in Table 4 it can be stated that KGIS potentially may deliver a high or very high support quality for nearly all elementary actions of the KW actions under consideration. An ever-growing stack of well-elaborated standards hosted by the W3C and other organizations of world-wide relevance builds a strong foundation for ongoing technological improvements. Another accelerating factor is the impressive growth of openly, standard-based accessible external knowledge sources, so-called Linked Open Data, which can be used to enrich substantially the internal knowledge base of an organization.

But there are also limiting factors and impediments for a broad implementation of KGIS. Five of them covering the aspects of infrastructure, technology, organization, and media effects will be discussed in the following:

7. *Persistence of established infrastructure*: Even if NoSQL databases begin to gain importance, particularly in big data applications, most of standard organizational information systems stick to the well-established, mature infrastructures based on relational databases.

8. *Media hype effects connected with upcoming technologies*: Since people in general tend to admire one-hit, preferably simple solutions to different, complex problems, media hype effects can be observed with nearly each upcoming technology. More than 10 years ago Semantic Web faced such a hype followed by a consequential "through of disillusionment". The Gartner Hype Cycle for Emerging Technologies [26] sees the "Enterprise Taxonomy and Ontology Management" even at the slope down with a perspective of more than 10 years to mainstream adoption. Furthermore, two other hypes disguise the relevance of Knowledge Engineering: Big Data and Machine Learning. It is often overlooked that these two technologies, as methods of structuring unstructured data, can seriously benefit from a conscious modeling of the domain's knowledge.

9. *Competence barriers in the community of web developers*: Web developers are faced with an impressive amount of new and hip technologies for designing and implementing small pieces of powerful software. They unfortunately refuse to gain additional competences in semantic technologies, except schema.org for standard typing of web content elements. But this technology is supported by big players of the Web industry and is oriented at the optimization of search engines, not KMIS.

10. *Indistinct role definitions for knowledge engineering*: Organizations in knowledge-intensive industries and in administration have started to recognize viable use cases for knowledge engineering. Nevertheless, they struggle to define distinct roles in this new area. Interestingly, the project teams encompass experts from very different areas in all cases observed by the authors.

11. *Week engineering support by editing tools*: Since knowledge engineering was most of the time driven by scientific enthusiasts, editing tools initially have been focused on full-fletched ontologies and reasoners. When first steps were taken to enter the business, they were quickly dismissed as too complicated. As a reaction, emerging tool developers started to build tools for very light-weight knowledge organization systems, so-called thesauruses focusing on weak broader-narrower relations between concepts without a distinct classification of objects and relations.

The authors are convinced that these impediments can be overcome with the help of joint projects of business, administration and applied sciences. To conclude the section the findings shall be illustrated by the main results of a small case study performed by [27] at a German higher education institution.

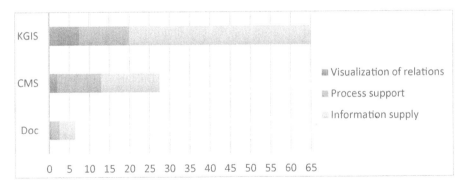

Fig. 5. Relative utilities (in %) of different implementation types of an IT service catalog

Case study "Semantic IT service catalog". Traditional IT service catalog implementations are mostly based on common CMS. A small number of public organizations uses document-based catalogs. Since none of these two implementation types meet all the valid requirements against an IT service catalog, the idea of a KGIS arises. A study was performed to prove the hypothesis that such a semantic system delivers a better support quality for KW actions. For getting reliable results in a mostly qualitative analysis, Analytical Hierarchy Process (AHP) [28] was chosen as methodology.

Based on a preliminary conceptual analysis, the main features of an IT service catalog were consolidated as follows: (i) visualization of relations, (ii) process support, (iii) information supply. As participants, a small number of domain experts and key users was involved. According to the AHP method they were asked – each independently – first to weight the selection criteria (the consolidated features) and then the implementation alternatives with respect to this criteria by pairwise comparison. The results are shown in Fig. 5. They prove the hypothesis given above.

6 Conclusion and Further Work

The presented paper pursues an integrative approach relating two scientific fields: (i) the organizational studies and (ii) the science of systems analysis and design with respect to information systems. From (i) are taken the concept of knowledge workers (KW) and their typical actions, whereas (ii) is applied to knowledge management information systems (KMIS) used to perform these actions by KWs. Although the share of KWs in organizations as well as their influence on the organizational success are ever growing, the IT support of their work remains suboptimal. Most of KMISs in use are built on relational databases and/or document oriented structures.

Analyzing the KMIS support for two important KW actions it was shown, that none of the traditional KMIS paradigms provide adequate support. A solution to solve this problem might be another KMIS paradigm based on knowledge graphs. The impediments of a broad implementation of such systems were discussed and in conclusion evaluated as surmountable by cooperative efforts.

The authors aim on the one hand to proceed this analytical research with the objective to develop a method for assessing the contribution of Semantic-Web-based Knowledge Engineering to the maturity of organizational knowledge management. On the other hand, the design science approach will be fostered and extended in cooperative projects with knowledge-intensive organizations and structures. The findings will be implemented in prototypes of knowledge-graph-based information systems. An emphasis will be put on weak but nevertheless crucial system elements, like editors for knowledge engineers and subject matter experts.

References

1. Davenport, T.H., Prusak, L.: Working Knowledge: How organizations manage what they know. Harvard Business School Press, Boston (2000)
2. Sajeva, S.: The analysis of key elements of socio-technical knowledge management system. In: Economics and Management, pp 765–774 (2010). ISSN 1822-6515
3. Hoq, M.G.K., Akter, R.: Knowledge management in Universities: role of knowledge workers. Bangladesh Journal of Library and Information Science 2(1), 92–102 (2012)
4. Mohanta, G.C., Kannan, V., Thooyamani, K.P.: Strategies for Improving Productivity of Knowledge Workers - An Overview. Strength Based Strategies, India (2006). http://strengthbasedstrategies.com/papers.htm
5. Laha, A.: A Theoretical Foundation for Building Knowledge-Work Support Systems. SETLabs, Hyderabad (2008). http://arxiv.org/abs/0910.5386
6. Drucker, P.F.: Managing in a Time of Great Change. Truman Talley Books/Dutton, New York (1995)
7. Reinhardt, W., Schmidt, B., Sloep, P., Drachsler, H.: Knowledge worker roles and actions – results of two empirical studies. Knowl. Process Manage. 18(3), 150–174 (2011)
8. Morello, D., Caldwell, F.: What Are Knowledge Workers? What Makes Them Thick? Gartner Group Research, Note SPA-12-7780 (2001)
9. Davenport, T.H.: Thinking for a Living: How to Get Better Performances and Results from Knowledge Workers. Harvard Business School Press, Boston (2005)
10. Figurska, I.: Knowledge workers engagement in work in theory and practice. Hum. Res. Manage. Ergonomic 2(9), 43–59 (2015)
11. Serrat, O.: Managing Knowledge Workers. Asian Development Bank, Washington (2008). http://digitalcommons.ilr.cornell.edu/intl/146/
12. Skrzypek, E.: Creativity of Knowledge Workers and Their Impact on Innovativeness of Enterprises (in Polish) (2009). http://www.instytut.info/Vkonf/site/32.pdf
13. Rizun, M.: Moodle platform as a knowledge management system: results of a questionnaire research. Econ. Stud. Sci. J. Univ. Econ. Katowice, Inf. Econometrics 296(6), 49–63 (2016)
14. Probst, G., Raub, S., Romhardt, K.: Wie Unternehmen ihre wertvollste Ressource optimal nutzen, 6th edn. Gabler Fachverlag, Wiesbaden (2010)
15. Maier, R., Hädrich, T.: Centralized versus peer-to-peer knowledge management systems. Mastering Knowl. Organ. Challenges Practices Prospects 13(1), 47–61 (2006)
16. Wikipedia Deutschland Analyse (2017). https://de.wikipedia.org/wiki/Analyse
17. Allemang, D., Hendler, J.: Semantic Web for the Working Ontologist: Effective Modeling in RDFS and OWL. Morgan Kaufmann, Amsterdam (2011)
18. Sowa, J.F.: Conceptual Structures: Information Processing in Mind and Machine. Addison-Wesley Longman Publishing, Boston (2008)

19. Stokman, F.N., de Vries, P.H.: Structuring knowledge in a graph. In: van der Veer, C.G., Mulder, G. (eds.) Human-Computer Interaction: Psychonomic Aspects, pp. 186–206. Springer, Heidelberg (1988). doi:10.1007/978-3-642-73402-1_12
20. Pujara, J., Miao, H., Getoor, L., Cohen, W.: Knowledge Graph Identification. In: Alani, H., et al. (eds.) ISWC 2013. LNCS, vol. 8218, pp. 542–557. Springer, Heidelberg (2013). doi:10.1007/978-3-642-41335-3_34
21. Blumauer, A.: From Taxonomies over Ontologies to Knowledge Graphs - The Semantic Puzzle (2014). https://blog.semantic-web.at/2014/07/15/
22. Miao, Q., Meng, Y., Zhang, B.: Chinese enterprise knowledge graph construction based on Linked Data. In: Proceedings of the 2015 IEEE 9th International Conference on Semantic Computing, IEEE ICSC 2015, Anaheim, California, pp. 153–154 (2015)
23. Galkin, M., Auer, S., Scerri, S.: Enterprise knowledge graphs: a survey. In: 37th International Conference on Information Systems (2016)
24. Paulheim, H.: Knowledge graph refinement: a survey of approaches and evaluation methods. In: Semantic Web, pp 1–20 (2016)
25. Meister, V.G., Jetschni, J., Kreideweiß, S.: Konzept und Prototyp einer dezentralen Wissensinfrastruktur zu Hochschuldaten für Mensch und Maschine (accapted to: Workshop Hochschule 2027 im Rahmen der 47. GI Jahrestagung, Chemnitz) (2017)
26. Gartner: Gartner's 2016 Hype Cycle for Emerging Technologies Identifies Three Key Trends That Organizations Must Track to Gain Competitive Advantage (2016). http://www.gartner.com/newsroom/id/3412017
27. Meister, V.G., Jetschni, J.: Umsetzungskonzepte und Nutzen von IT-Dienste-Katalogen für die IT-Versorgung in Organisationen – Eine Fallstudie im Hochschulumfeld. In: Angewandte Forschung in der Wirtschaftsinformatik. 28. Jahrestagung des AKWI, Luzern, pp 141–151 (2015)
28. Saaty, T.L., Vargas, L.G.: Models, Methods, Concepts & Applications of the Analytic Hierarchy Process. Springer, New York (2012). doi:10.1007/978-1-4615-1665-1

Web-Based Information Systems

Comparative Analysis of e-Banking Services in Poland in 2016

Witold Chmielarz and Marek Zborowski[✉]

Faculty of Management, University of Warsaw,
Szturmowa 1/3 Street, 02-678 Warsaw, Poland
witold@chmielarz.eu, mzborowski@wz.uw.edu.pl
http://www.wz.uw.edu.pl/en/pracownicy/lista/witold-chmielarz
http://www.wz.uw.edu.pl/en/pracownicy/lista/marek-zborowski

Abstract. The objective of this article is to identify the best e-banking services in the most popular banks in Poland in 2016 from the point of view of an individual client. Electronic banking leads to strengthening the position of a particular bank in the competitive market environment. Thus, the high quality of its website frequently plays a major role regarding the perception of the bank. This paper focuses on the ways to secure and strengthen the position of a bank in the sector using its website to enhance its image, improve usability and communication with clients. Following a brief introduction describing the situation of electronic banking in Poland, the authors present the assumptions adopted for the conducted research. Subsequently, on the basis of the obtained findings, the authors have carried out multidimensional analyses and presented the resultant conclusions and recommendations. The authors original contribution was specifying the criteria used for websites evaluation and applying the conversion method constructed by them for conducting such analyses.

Keywords: e-Banking · e-Services · Evaluation of web sites

1 Introduction

The pace of the development of e-banking services in Poland may constitute a model for other sectors of the economy. Compared to the second quarter of 2015, the number of individual clients with potential access to accounts increased by nearly 13% in relation to the second quarter of 2016, reaching the value of 31.509 million users; out of which, the number of active individual clients increased by almost 8%, reaching the level of 15.200 million [13]. This is undoubtedly the fastest growing banking sector, and nothing points to the fact that the positive trend might be changed. Every year the population of new clients using the opportunities offered by the Internet to handle banking transactions is growing. Among all the clients having electronic access to account, there are over 48% of active users of e-banking.

© Springer International Publishing AG 2017
S. Wrycza and J. Maślankowski (Eds.): SIGSAND/PLAIS 2017, LNBIP 300, pp. 43–55, 2017.
DOI: 10.1007/978-3-319-66996-0_4

The problems connected with the evaluation of websites, in particular, access to e-banking services are widely discussed in literature; however, there is no single solution which would prevent them. The literature review [1,3,6,10–12,15,16] shows that e-banking websites may be analysed from the point of view of:

- usability,
- functionality (search, navigation, the relevance of content),
- interactivity (availability and responsiveness),
- visualization (color scheme, background, graphics, text),
- reliability,
- effectiveness.

Most of the methods applied in the evaluation of e-banking websites are traditional scoring methods based on specific sets of criteria, assessed according to a standardized scale. Technical and functional criteria are usually among the factors which appear most frequently. Many of the considered factors may be evaluated in a very subjective way: text clarity, attractive color scheme, images and photos, the speed and intuitiveness of navigation), etc. What is more, some users do not treat the particular criteria sets in an equivalent way. On the other hand, there occur numerous problems with determining particular preferences and the evaluation of relations between them. This type of comparative analysis is carried out in three major cases which allow:

- specification and accurate examination of the area where particular software is applied,
- construction of the ranking of IT solutions existing in the market,
- identification of the qualities for creating recommendations for websites analysis and design.

The presented studies focus on the first and the second of the cases mentioned above. So the main problem being analyzed is the specification of the determinants influencing the evaluation of e-banking services by an individual client in Polish banks and showing which ones are the most important.

2 The Population Sample and the Research Method Analysis

In June 2016, the authors carried out studies into the quality of e-banking websites of the banks which are most popular among individual clients in Poland. The sample included 146 respondents. The research was the case of purposive sampling – the study was conducted by electronic questionnaires among the students of the last years of accounting, finance and Insurance specialization studies at Faculty of Management University of Warsaw, in the age of 19–45, in randomly selected student groups. Among the survey participants, there were 74% of women and 26% of men, mainly from Warsaw and surrounding areas. Each respondent declared having at least one electronic access to an account, in at least one of the banks operating in the territory of Poland (21 participants to two accounts, and 6 to three). It means that even then when 135 people filled the questionnaire correctly, thus, the present study examined the data concerning 168 active accounts with access to e-banking services.

The biggest number of internet access accounts were indicated in the case of mBank clients (24.39%), next position was taken by Inteligo PKO BP S.A. (18.29%) and ING Bank Slaski (14.63%). The smallest number of users in the examined sample declared having accounts in: Bank Pocztowy (2.44%), Credit Agricole (2.44%) and BPH (3.66%). Out of the first ten bank most popular amongst respondent were eg. Deutsche Bank or Toyota Bank.

The calculations presented in this paper were carried out with the application of the authors' own set of criteria, established based on the literature and consultations with experts, used by the authors since 2009, applied to evaluate the electronic access to the services of selected banks. The criteria applied in the study may be divided into two basic groups:

– economical – annual nominal fee, a fee for maintaining an account PLN/month, a surcharge for access to electronic channels, a fee for a transfer to a parent bank, a fee for a transfer to another bank, interest rate on deposits – a deposit of PLN 10,000, a fee for issuing a card and a monthly fee for a card – PLN/month,
– technical – functional – due to considerable similarity of basic services the authors only selected non-standard additional services such as: insurance, investment funds, cross-border transfer or foreign currency account; technological – the number of surcharge-free ATMs, quantites of account access channels; and security.

At the time of the economic crisis, the criteria used to evaluate the websites offering access to e-banking services as presented above, were supplied with the set of psychological criteria, including the so-called anti-crisis measures which comprised all – as regarded by the consultants cooperating with the authors – manifestations of activity undertaken to counteract the potential effects of the crisis in e-banking sector [8]. The set of these factors has been presented for consideration of interviewees during the most recent evaluation of e-banking websites. The proposed set of anti-crisis measures included:

– the dynamics of interest rates on deposits,
– the dynamics of interest rates on credits,
– stability of the bank policies concerning basic charges,
– the level of clients' trust,
– the average of positions occupied in internet rankings.

The presented study is the latest in the series of cyclical rankings whose basic aim is to evaluate the factors impacting the usability of the websites offering access to individual accounts in banks. The same sets of evaluation criteria were applied in the analyses of the condition and the changes in electronic banking in 2013, 2014 and 2015. They were originally established during an internet discussion, initiated by the authors, involving the researchers from reputable universities carrying out research into the issue of electronic banking in Poland. In order to evaluate particular criteria in the banks selected by students, the authors used a simplified, standardized Likert scale [9]. The study was carried out with the application of a simple scoring method. In the simple scoring method the authors measured the distance from the maximum score to be obtained. This concerns the value of criterion measure and in the sense of the distance it is the same when we measure the distance from the first criterion to the second one, and the other way round. However, we do not define the relations between particular criteria. Scoring methods are regarded as subjective, though its subjectivity appears to be reduced with increasing the number of respondents in the research sample. AHP method (T. Saaty), Promethee II, Electre I and III, and other similar methods are believed to be more objective. The experience of the authors [4, 6, 17], primarily related to the application of AHP method in the websites' evaluation shows, however, that from the point of view of the respondents completing the survey questionnaires proves very difficult. As a result, it frequently leads to illconsidered and accidental judgments, and the score is often affected by the order in which particular criteria appear. To eliminate the above-described problems, the authors have developed their own author's evaluation method – a conversion method which combines simplicity and unambiguity of the scoring method with the precision of relational methods (full description see [5]). This method consists in determining the relation of each criterion to other criteria, based on averaged distances from the maximum potential value established on the basis of previous scoring evaluation. Data received from scoring evaluation is the starting point for a conversion method. Then, we adopt the following assumptions: after constructing the experts' table of evaluations of particular criteria for each website, we need to perform the conversion with the established preference vector of the superior level criteria. Next, the authors perform the transformation of the combined scoring table into the preference vector (first converter).

The next steps are:

– constructing a matrix of distances from the maximum value for each criterion in every website, establishing the maximum value:

$$P_{i,max} = Max\{f_i(a_j), ..., f_n(a_m)\}$$

for

$$i = 1, ..., n$$

and

$$j = 1, ..., m;$$

– establishing the matrix of the distances from the maximum value

$$\delta(f_j(aj) = P_{i,max} - f_i(a_j)$$

for

$$i = 1, ..., n$$

and
$$j = 1, ..., m;$$

– calculating the average distance from the maximum value for each criterion,

$$\overline{F_{i,j}} = \frac{\sum_{jm1}^{m} \delta(f_i(a_j))}{m}$$

– as a result of the above operation, constructing a matrix of differences in the distance from the maximum value and the average distance according to criteria,
– for each bank website: constructing conversion matrices - modules of relative distances of particular criteria to remaining criteria (the distance from the same criterion is 0), the obtained distances below the diagonal are the converse of the values over the diagonal,
– averaging criteria conversion matrices - creating one matrix of average modules of values for all criteria:

$$\overline{A_{i,j}} = \frac{\sum_{i=1,j=1}^{n,m}(\alpha_{i,j} - \alpha_{i+2,j})}{n}$$

– transforming the conversion matrix of criteria into a superior preference matrix (calculating squared matrix, adding up rows, standardization of the obtained preference vector; repeated squaring, adding up rows, standardization of preference vector - repeating this iteration until there are minimum differences in subsequent preference vectors).

As a result of the above operations we establish a criteria conversion matrix Ta_{mx1}. Subsequently, the authors performed a transformation of the scores presented by experts on the level of a matrix specifying expert websites' evaluations for particular criteria (second converter). The results have been obtained in an analogical way:

– constructing a matrix of distances from the maximum value for each criterion and each website:
   • establishing the maximum value

$$P_{i,max} = Max\{f_i(a_j), ..., f_n(a_m)\}$$

for
$$i = 1, ..., n$$

and
$$j = 1, ..., m;$$

   • establishing the matrix of distances from the maximum value

$$\delta(f_i(a_j)) = P_{i,max} - f_i(a_j)$$

for
$$i = 1, ..., n$$

and

$$j = 1, ..., m;$$

- calculating the average distance from the maximum value for each website,

$$\overline{F_i} = \frac{\sum_{j=1}^{m} \delta(f_i(a_j))}{m}$$

– constructing a matrix of the differences of deviations from the maximum value and the average distance of the features from the maximum,
– for each criterion: constructing a matrix of transformations (conversions) of the differences of the average distance from the maximum value between the websites, analogically as presented above values below the diagonal are the converse of the values over the diagonal,
– constructing a module matrix of transformations of the differences of average distance from the maximum value between the websites, for each criterion,

$$\overline{A_{i,j}} = \frac{\sum_{i=1,j=1}^{n,m}(\alpha_{i,j} - \alpha_{i+2,j})}{n}$$

– for each module matrix of transformation of the differences of the average distance from the maximum value between the websites, squaring it, adding up rows, standardization of the obtained ranking vector and repeating this operation until the obtained differences between two ranking vectors for each criterion will be minimal,

As a result of the above presented operations we obtain a conversion matrix of websites' evaluations: Tf_{mx1}

– using the obtained vectors to construct a combined ranking matrix - returning to the matrix where in
– its side-heading there are criteria, in the heading names of bank websites by appropriate transfer of the obtained preference vectors for each criterion,
– multiplying the matrix obtained in such a way by the previously calculated preference vector,

$$T' = Tf \otimes Ta$$

– analysing final results and drawing conclusions (Note: the lowest distances in this case are the most favourable, comparability adjustments to other methods can be obtained by subtracting these values from 1 and their repeated standardization).

The basis for the creation of the presented method was the assumption that it should be easy to apply. The objective has been reached, which is visible in the number of the advantages presented below. The only disadvantage of the method is the fact that the transformation of the results of the survey is connected with carrying out many complex operations. The advantages of this method are:

- the ease of application (similar to the implementation of the scoring method) which results from the fact that in the survey, analogically to the scoring method, there are questions concerning the subjective evaluation of the element,
- in the case of considering a large number of evaluation criteria, or alternatives there is no significant increase in the number of questions contained in the survey, necessary to consider (as in the AHP method),
- the possibility of the application of the method in the studies conducted with the participation of respondents who are not experts in the field,
- there are no measures, as in the case of ELECTRE method - veto threshold, which may be not fully understandable for the respondents [2],
- the result of the calculations is relatively easy to interpret since it takes the form of the rank of the evaluations of the examined objects.

The above mentioned induced the authors to apply it to verify the subjectivity of the evaluations carried out by means of a scoring method.

3   Analysis of the Findings and Discussion

To evaluate economical, technical and anti-crisis criteria, the authors used a preliminary table presenting bank offers related to e-banking services used by the respondents and fees connected with using bank accounts operated via the Internet, created on the basis of data obtained from websites of particular banks. On the basis of the completed questionnaires, the authors created an averaged combined table of the criteria generated by the users. The best in the present classification are: Credit Agricole (79.86%) and T-Mobile Uslugi Bankowe (79.44%). Immediately behind are: Bank Pocztowy and ING Bank Slaski. Interestingly, the second position is taken by a mobile bank which has been established as a result of cooperation between the most innovative Alior Bank (taking the 11th position in our ranking) and the largest mobile operator T-Mobile, basing upon the experience of Alior Sync. Remarkably low position – sixth place – was taken by mBank, which so far was a leader in the ranking and was hugely popular in the analyzed group of respondents (interestingly, the low scores were assigned for: functionality, clarity and user-friendliness and anti-crisis measures – 0.68. In the rankings carried out until May 2010 the bank maintained its first position. The second conclusion, which seems to be characteristic of this study – the overall scores for the quality of websites tend to increase. The worst in the ranking were: Bank BPH and Getin Bank. The first four banks in the classification were placed above the average amounting to 75.50%. The results of the ranking are presented in Fig. 1.

In the case of most banks examined in the study, there are no additional charges for issuing a debit card, and transfers to a parent bank are usually free of charge. The level of security may be seen as sufficient for the clients. Basically, this has not changed since 2008. The summary table shows that the fee for issuing a card (usually, no charge), reached the level which at present is satisfactory for the client in 100% (96%). The satisfaction related to the fee for a transfer to

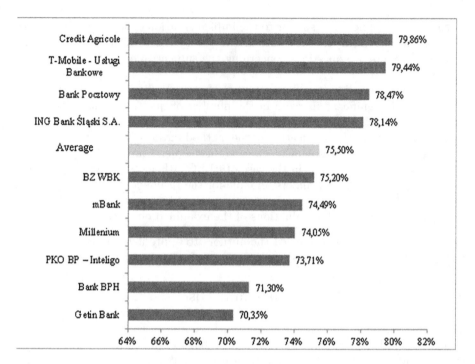

Fig. 1. Ranking of the quality of e-banking websites of selected banks in Poland in 2016 according to the scoring method

a parent bank is usually also very high (over 96%). Undoubtedly, the worst indicator in the ranking is annual nominal interest rate (in most cases assessed by the users as too low – 34.20% of the maximum score). In fact, we may see that the savings accounts which tend to appear are a specific response to the decrease in annual nominal interest of accounts. Over 91% of the maximum scores were obtained by the fee to another bank. Following the recent implementation of the government regulations, most of the economical indicators were placed below the users' average assessment of 75.5% (Fig. 2).

Among the factors which were not listed within the criteria, the clients pointed to the lack of possibility to make a cross-border transfer (e.g. SWIFT in Inteligo) or no possibility of fully automatic obtaining a credit. In 2008 there were no anti-crisis measures among the criteria – if we compare the study with the research carried out in 2014, we need to admit that during the crisis e-banking clients did not notice any signs of the crisis and they were not able to define the anti-crisis measures undertaken by the banks, and at present they sometimes suggest the relevant criteria to be applied for their evaluation. The first three banks in the ranking this year are new players in the e-banking market, which in the last three years ranked below the top ten positions in the rankings. The present-day leaders owe their present position to banking applications running on smartphones and tablets. However, generally, in recent years the highest scores

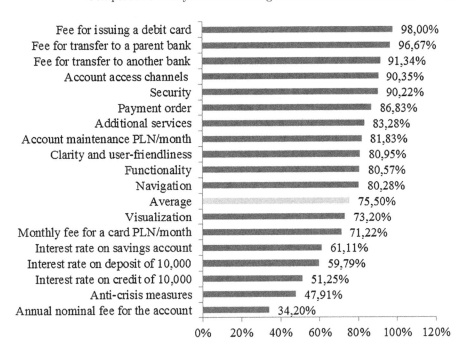

Fig. 2. Ranking of the criteria used for the evaluation of electronic access to individual accounts in selected banks in Poland in 2016 according to the scoring method

were awarded to established banks with a strong position in traditional internet banking, such as ING Bank Śląski S.A., Bank BPH or BZ WBK. Among the top ten positions, Millenium and Getin Bank held their positions, as well as the banks which were pioneers with regard to electronic banking services in Poland and have a large group of loyal customers, especially middle-aged clients.

In the present study, the applied conversion method produced interesting results. Due to a large degree of discrepancy with reference to the users' opinions on the e-banking services in the same banks, not only it averaged and reduced the differences, but also, taking into account the relationship between the maximum and average values obtained as a result of calculations, it caused a significant change in the scores. Previously, a similar effect was achieved using the Saaty's AHP method. "Flattening" the extreme opinions allowed the authors to obtain results closer to the opinions of the most active clients than in the case of the scoring method. In the case of the classification carried out with the application of the conversion method, the top positions were taken by the banks which pursue the most stable policy from the perspective of the clients. Thus, these were not the services which relatively recently emerged in the market, since the users' opinions concerning their activity are still largely unstable (a wide spread of responses). Here, the results were markedly lower for the banks which in the scoring evaluation ranked higher.

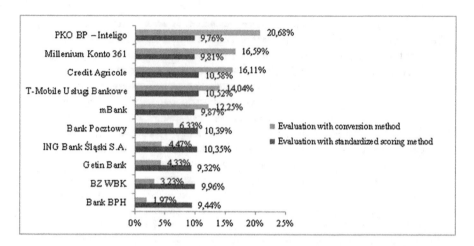

Fig. 3. Comparison of the evaluations carried out with the scoring method and the conversion method for ten most popular banks in Poland in 2016

It appears to be the result, on the one hand, of the increased awareness of the clients; on the other, this was due to the fact that in the case of "their" e-banking websites the changes (not always seen as improvements) were too rarely introduced. Also, the results of the ranking have changed, sometimes to a considerable degree. PKO Inteligo was a leader in the ranking, moving from the previous eighth position; the second place was taken by Bank Millenium (from the seventh position), and the subsequent positions were taken by the banks which took leading positions in the "scoring" ranking: Credit Agricole, T-Mobile Uslugi Bankowe and Bank Pocztowy. BZ WBK fell by four positions. Bank BPH, similarly to previous rankings, took the last position (Fig. 3).

The position of mBank appears to be the most stable one – the difference between the scores slightly exceeds two percentage points. In the evaluation carried out with the conversion method, the discrepancy between the scores amounted to nearly 19%, in the standardized scoring method it was about 20 times lower (the difference of 9.5 of the percentage point was indicated in the case of the share in the maximum scores of particular criteria).

4   Conclusions

The presented analysis has indicated the great diversification of the opinions of individual clients on the issue of the application of e-banking systems, particularly the views concerning the selection and use of websites to deal with everyday e-banking transactions and operations. Despite the 80% of available e-banking services and above 40% of clients who are actively using their accounts, it is estimated that further development of this sector in Poland has considerable potential.

If we relate the present findings to the results of previous studies [4], we may note that the analysis indicates the changes taking place in recent years was regards the clients' awareness and activity. An individual client of e-banking services changes from the user of its simplest functions into an experienced user who can point to advantages and disadvantages of this modern form of communication. Also, the consumers are able to recognize the benefits obtained from using an electronic access account and show where they can obtain the greatest benefits. The results are the choices made by the clients reflected in the presented study and commented on in the survey. To recapitulate – they lead to the following conclusions:

- the most significant phenomenon occurring in the electronic banking market is mobile access to banking services. It takes two forms: access via banking websites and by means of banking applications on the devices such as smartphones and tablets. Banks which offer access to Windows websites via mobile devices start to move to leading positions in all the rankings. Surprisingly, T-Mobile Usługi Bankowe (founded on the basis of Sync Bank) receive much better scores than the multi-channel partnership, Alior Bank,
- in 2016 most banks operating in Poland experienced a wave of increases caused by rapidly falling interest rates and the externally imposed statutory reduced interchange fees levied on trade. In 2017 we may expect next increases resulting from both low interest rates and bank tax (0.44% of assets) and higher contributions to the Bank Guarantee Fund (Bankowy Fundusz Gwarancyjny) (expenditure allocated to bailing out bankrupt credit and savings unions (SKOK)). Unfortunately, the specific inactivity of clients, result in low mobility in relation to changes introduced by particular banks,
- further increases which are to be expected in 2017 may cause major changes in the rankings due to their almost universal character. In 2016 there occurred an increase in the account maintenance fee, cards and cash withdrawals from ATMs, e.g. ING, Pekao. At present, many banks plan to introduce changes in their pricing policies: Bank Handlowy (account maintenance, using a debit card, payment order and ATM withdrawals), Citi Bank (using cards, selected charges for transfers, withdrawals from ATM of another bank, an increase of fee in the case of not using accounts, ING Bank – changes in the terms concerning deposits, withdrawals, transfers in the bank branch, etc. The unclear and complicated way of introducing the changes once again may delay the transformations in the consumer market, but perhaps the actual shifts may finally take place,
- since few customers dynamically react to changes of basic evaluation criteria for banking services. Their reasons for the lack of mobility are as follows: a confusing and complicated structure of banking services; changes implemented without informing clients (banks tend to believe that it is sufficient to post information on a website etc.); eventually, habit and resistance to changes which require time and patience to arrange for and handle the changes),
- Bank Credit Agricole is popular due to intuitive communication both with regard to traditional contacts – using a browser and mobile access with a smartphone or a tablet, as well as, at the time of the survey, reasonable and competitive terms of using the account and other e-banking tools (cards),

– T-Mobile Uslugi Bankowe is a leader in the market not only due to the mobile access to services, but also other attractive terms e.g. the fact that it pays interest on the funds regularly deposited in a personal account, and offers withdrawals from all ATMS in Poland and abroad without additional charges,

– vast majority of active bank customers regard economical factors as the most important criteria for the evaluation of electronic access to banking services – usually the prices of the most frequently used services. The prices, however, tend to be more and more similar, even the banks with the highest scores in the ranking, as a result of amended governmental regulations and the behavior of competitors establish e.g. "punitive" card charges for the clients who do not use it to a "sufficient" degree,

– more and more people, however, admit that when selecting a website they tend to focus more on the user-friendliness of the website, its visual attractiveness or unlimited access to the account in any place and at any time (e.g. using m-banking application). As far as basic account operations are concerned, they even speak of the substitutability of these criteria,

– some customers pay attention to the redundancy of banking services offered by websites in relation to the most essential needs, lack of tailor-made offers, intrusive advertising which appears to be most profitable for the bank, not for the clients, and on the other hand, limiting such redundancy not only in mobile applications, but also mobile device access to Windows systems of these banks,

– the number of non-active users is worryingly high in relation to those who may potentially use e-banking services. Until a few years ago, these estimates did not exceed 20%, which indicates a deliberate action of banks towards "pushing" products rather than meeting their customers' needs. Such a phenomenon could be observed previously in the case of using payment cards and – with time this percentage started to decrease, which has not occurred in relation to internet or mobile access to electronic account.

The diversity and dynamics of the evaluations is important for practice - confirm the thesis concerning the necessity to regularly examine this sphere in the matter of the clients' use of e-banking services and the tendencies concerning designing websites which are characterized by high usability from the clients' point of view. It also points to the need for further studies aimed at constructing a multi-dimensional, multi-criteria, hierarchical and multi-faceted system of the evaluation of websites, with the consideration of additional, more specific criteria such as e.g. customer profile [7], which, so far, has not been considered to a sufficient degree. And of course making the survey on a larger sample of bank customers, because we are aware that the selected, specific sample is one of the weaknesses of this study.

One may observe that mobile account access is becoming a more and more important channel, and it takes the place previously taken by traditional access to an account with personal and desktop computers. Undoubtedly, this development irrevocably changes clients' expectations, perceptions and habits related to using banking services, and also simultaneously – it urges the banks to introduce quick changes of the medium which would take into account users' requirements.

References

1. Bauer, H.H., Hammerschmidt, M., Falk, T.: Measuring the quality of e-banking portals. International Journal of Bank Marketing **23**(2), 153–175 (2005)
2. Buchanan, J., Sheppard, P., Lamsade, D.V.: Project ranking using ELECTRE III. http://130.217.168.130/departments/staff/jtb/Electwp.pdf. Accessed Jan 2015
3. Chiou, W.C., Lin, C.C., Perng, C.: A strategic framework for website evaluation based on a review of the literature from 1995–2006. Information & Management **47**(5–6), 282–290 (2010)
4. Chmielarz, W., Zborowski, M.: Comparative analysis of electronic banking websites in Poland in 2014 and 2015. In: Ziemba, E. (ed.) Information Technology for Management. LNBIP, vol. 243, pp. 147–161. Springer, Cham (2016). doi:10.1007/978-3-319-30528-8_9
5. Chmielarz, W., Zborowski, M.: Conversion method in comparative analysis of e-banking services in Poland. In: Kobyliński, A., Sobczak, A. (eds.) BIR 2013. LNBIP, vol. 158, pp. 227–240. Springer, Heidelberg (2013). doi:10.1007/978-3-642-40823-6_18
6. Chmielarz, W., Szumski, O., Zborowski, M.: Kompleksowe metody ewaluacji jakości serwisów internetowych. Wydawnictwo Naukowe WZ UW, Warsaw (2011)
7. Chmielarz, W.: Methodological aspects of the evaluation of individual E-Usługi Bankowe for selected banks in Poland. In: Pańkowska, M. (ed.) Infonomics for Distributed Business and Decision-Making Environments. Creating Information System Ecology. IGI Global, Business Science Reference, Hershey-New York (2010)
8. Cyfrowa Polska: Szansa na technologiczny skok do globalnej pierwszej ligi gospodarczej, McKinsey, Forbes Polska (2016). www.mc.kinsey.pl, www.forbes.pl
9. Likert, R.: A technique for the measurement of attitudes. Arch. Psychol. **140**, 1–55 (1932)
10. Mateos, M.B., Mera, A.C., Gonzales, F.J., Lopez, O.R.: A new Web assessment index: Spanish universities analysis. Internet Res. Electron. Appl. Policy **11**(3), 226–234 (2001)
11. Migdadi, Y.K.: Quantitative evaluation of the internet banking service encounter's quality: comparative study between Jordan and the UK Retail Banks. J. Internet Banking Commer. **2**(13), 1 (2008)
12. Miranda, F.J., Cortes, R., Barriuso, C.: Quantitative evaluation of e-banking web sites: an empirical study of Spanish Banks. Electron. J. Inf. Syst. Eval. **2**(9) (2004). http://www.eiise.com
13. NETB@nk Raport Bankowość internetowa i płatności bezgotówkowe. Podsumowanie II kwartału 2016 r., Związek Banków Polskich (The Polish Bank Association) (2016). http://www.zbp.pl/Netbank_Q2_20160927.pdf. Accessed 15 Dec 2016
14. Webb, H.W., Webb, L.A.: SiteQual: an integrated measure of web site quality. J. Enterp. Inf. Manage. **17**(6), 430–444 (2004)
15. Wielki, J.: Modele wpływu przestrzeni elektronicznej na organizacje gospodarcze. Wydawnictwo UE we Wrocławiu, Wrocław (2012)
16. Yang, Z., Cai, S., Zhou, Z., Zhou, N.: Development and validation of an instrument to measure user perceived service quality of information presenting Web Portals. Inf. Manage. **42**(4), 575–589 (2005)
17. Zborowski, M.: Modelowanie witryn internetowych uczelni wyższych o profilu ekonomicznym, Faculty of Management, University of Warsaw, doctoral dissertation (2013)

Integration of Eye-Tracking Based Studies into e-Commerce Websites Evaluation Process with eQual and TOPSIS Methods

Jarosław Wątróbski[1(✉)], Jarosław Jankowski[1], Artur Karczmarczyk[1], and Paweł Ziemba[2]

[1] West Pomeranian University of Technology in Szczecin,
ul. Żołnierska 49, 71-210 Szczecin, Poland
{jwatrobski,jjankowski,akarczmarczyk}@wi.zut.edu.pl
[2] Department of Technology, The Jacob of Paradies University, ul. Teatralna 25,
66-400 Gorzów Wielkopolski, Poland
pziemba@ajp.edu.pl

Abstract. The evaluation of the e-commerce websites' quality, usability and user experience is an important research task. Various survey-based evaluation methods are available, eQual being one of them. However, these methods are not free of disadvantages. In this paper, a novel approach is presented, where a single usability evaluation model is created on the basis of the eQual survey criteria and the perceptual evaluation data from eye tracking devices, thus extending the statistical survey model with objective gaze measurements and Multi-Criteria Decision Analysis (MCDA) methodology foundations. The combined data is processed with the crisp and fuzzy variants of the TOPSIS method to evaluate the websites. An empirical verification is performed and the results are presented. The results showed the benefits of the author's proposed approach as well as the wide possibilities of interpretation of the obtained solutions.

Keywords: TOPSIS · eQual · Eye tracking · Websites quality evaluation

1 Introduction

In January 2017, 3.77 billion people out of the total world population of 7.5 billion people were Internet users and over 4.9 billion people used mobile devices. A growth of 482 million users, i.e. 21%, was observed among active social media users since January 2016 [35]. According to January 2016 data [34], 71% of USA citizens searched online for a product or service to buy within a 30-day span period and 69% visited an online retail store. According to [40], the online sales in the US will reach \$523 billion by 2020.

Harsh competition can be observed in the e-commerce sector. In June 2016 only 5.4% of online stores sold more than \$1, 000 per year [51]. With such a heated up competition, entrepreneurs try to increase their chances by marketing and using analytic tools [58], refactoring the usability of the website and its

© Springer International Publishing AG 2017
S. Wrycza and J. Maślankowski (Eds.): SIGSAND/PLAIS 2017, LNBIP 300, pp. 56–80, 2017.
DOI: 10.1007/978-3-319-66996-0_5

assessment [25], providing web content accessibility [56] or building the credibility of the website by, among other things, providing high levels of usability and quality and providing best user experience [47].

As a result, the evaluation of the websites' quality, usability and user experience becomes a very important matter to the business owners [37]. The evaluation can be performed for example by the means of competitive analysis, scenarios, inspection methods, log analysis or on-line questionnaires [22]. There are multiple website and e-commerce evaluation methods based on the latter option, including eQual [11], SiteQual [66] or E-S-QUAL [49] to name just a few. These methods have been successfully used in the evaluation of the e-commerce [5], e-government [10], e-banking [49], information services [67] or university websites [8], however, they have some disadvantages. Firstly, to ascertain the responses are not artificial, the questionnaires need to be directed to an appropriate group of real users. Secondly, the list of questions must be short or the number of alternatives limited, otherwise the total number of questions would be too long for the respondents to complete. Also the reliability of the collected data needs to be verified. Last, but not least, possible alternatives ordering or comparison bias needs to be considered [11].

Furthermore, the websites' quality, usability and user experience evaluation problem encourages a multi-faceted viewpoint. Therefore, the Multi-Criteria Decision Analysis (MCDA) methods can be used. These include inter alia TOPSIS [31] or PROMETHEE [17] in their classic or fuzzy variants [14,59]. The MCDA methods give a profound methodological foundations for the multi-aspect data aggregation. Additionally, according to the recent research, the aggregation simplification that happens during the usage of the classic survey evaluation methods is less beneficial than using the MCDA based mathematical solutions [62].

This paper addresses the research gap of the lack of an integrated approach connecting the classical survey methods with eye tracking data, using dedicated mathematical approach for data aggregation. The authors' contribution is to create a unique integrated framework that facilitates the aggregation of data from multiple sources using a dedicated mathematical approach with the MCDA methodology foundations. In practical terms, the eQual method website quality evaluation criteria is extended with 6 perceptual metrics collected from an eye tracking (ET) device and selected e-commerce websites are evaluated in the empirical study.

The paper is split into sections. Section 2 contains literature review. The methodological framework of the proposed approach is presented in Sect. 3. Section 4 contains the empirical study results. The conclusions and future directions are outlined in Sect. 5

2   Literature Review

Nowadays, when the competitors are literally just a few clicks away, fine-tuning the e-commerce offering to the customer has become a key aspect of maintaining a profitable online business [47]. The authors of [42,64] note that the average human's focus time span is equal to eight seconds. For that reason, the websites need to be created in a manner that will allow customers to almost immediately find what they are looking for. This aim can be partially accomplished by achieving high levels of the site's quality, usability and user experience. It is important to note, however, that as the software, hardware and user preferences evolve over time, the usability of the systems changes as well [46]. Consequently, it is important to evaluate the quality, usability and user experience of e-commerce websites [37]. There are numerous ways to evaluate the quality of a Web site, such as competitive analysis, simulation of real users, comparison of the site interface with a predetermined set of guidelines, log analysis, surveys and on-line questionnaires to name just a few [22].

The surveys option is a very common choice for the quality evaluation. The procedure of collecting the participants' responses can be performed either locally, on a designated workstations as in [15], or on-line with the usage of Internet-based questionnaires, as in [9]. During the experiment, the participants express their preferences and opinions by answering multiple-choice questions using a Likert scale. Depending on the method, a 5-point, 7-point or similar scale is used, where low values express a strong disagreement and high values imply a strong agreement [41]. Subsequently, the obtained data is aggregated and verified. Eventually, the data analysis is performed using a variety of statistical methods.

There are multiple classic methods for performing such survey-based evaluations. The eQual method [10,11] is based on the Quality Function Deployment and uses 22 criteria assessed on a 7-point scale. It has been successfully applied inter alia in the evaluations of the e-commerce, e-government, university and WAP websites. It is worth mentioning that in eQual method not only the partial evaluations of each criteria but also the weights of the criteria are determined with the usage of questionnaires. The Ahn method, deriving from the Technology Acceptance Model and the Model of Information Systems Success, uses 54 criteria and was used to evaluate such systems as e-banking and e-commerce [5]. The Site-Qual method [65] was built on the SERVQUAL [50] and Data Quality. Contrary to the aforementioned methods, SiteQual uses a 9-point Likert scale in the questionnaires. The WEQ method [24] was applied for e-government evaluation. It is based on Website User Satisfaction model. Two sets of criteria are utilized – 18 positive and 8 negative. The respondents answer the questions on a 5-point scale. The WPSQ method [67] has the same theoretical basis as the Ahn method, but it uses 19 criteria and a 5-point answers' scale. The E-S-QUAL/RecS-Qual methods [6,49], which were successfully applied to the evaluation of e-banking and e-commerce websites derive from SERVQUAL and use 33 criteria. All the methods mentioned above require at least 30 evaluators.

Based on the number of citations and the wide range of applications, the eQual method can be considered to be one of the most popular evaluation methods. Supplementary to the evaluation based on individual criteria, each participant also provides an overall evaluation of the website, which is later used for the reliability verification of the questionnaire [12]. Subsequently, when the data from all surveys is collected, Cronbach's Alpha is employed to determine the reliability and cohesion of the results [13]. Eventually, the Evaluation Quality Index (EQI) is calculated on the basis of formulas 1–4 to obtain the final evaluation of the subject of analysis.

$$EQI = \sum_{k=1}^{m} EQI_k/m \tag{1}$$

$$EQI_k = (Score_k/Max_k) \cdot 100 \tag{2}$$

$$Score_k = \sum_{i=1}^{n} (o_i(k) \cdot w_i(k))/n \tag{3}$$

$$Max_k = \sum_{i=1}^{n} (7 \cdot w_i(k))/n \tag{4}$$

where m denotes the number of criteria, n – the number of polled users, $o_i(k)$ – the i-th user evaluation of the website considering the n-th criterion and $w_i(k)$ – the weight of the k-th criterion elicited from the i-th user.

The classic survey methods have been successfully used for evaluations in multiple disciplinary domains, nevertheless, there are some problems related to them. First of all, the reliability of the data retrieved needs to be verified. It can be achieved by using the Cronbach's Alpha mentioned above or some complex reliability tests such as convergence evaluation or discriminant analysis [65]. Additionally, the responses are usually limited to a n-point Likert scale. As the authors of [41] note, if too few rating categories are used, much of the raters' discriminative power is lost. Conversely, the scale can be graded so finely that it would be beyond the raters' capability of discrimination. Last, but not least, the issue of evaluation in some domains, such as the websites' user experience, quality and usability, encourages a multi-aspect approach [60,61].

Many multi-criteria decision analysis (MCDA) techniques provide synthetic measures which preserve the methodological quality of data aggregation inter alia by taking into consideration the weights assigned to each criterion, analyzing the implication of each weight or by the use of compensatory or non-compensatory strategies [26]. It is also worth noting that the MCDA methods, especially the ones originating from the American school, use advanced mathematical methods to produce the value/utility score to facilitate the evaluation, as opposed to basic operations utilized in the classic survey methods [45,54]. The Technique for order performance by similarity to ideal solution (TOPSIS), first developed by Hwang and Yoon [30] for solving a MCDA decision-making problem, allows to see the ideal and anti-ideal solutions and then compare the relative distances between them and the analyzed variants, based on the concept that the chosen alternative should be as close as possible to the positive ideal solution and as far as possible from the negative ideal solution.

As the MCDA methods have been developed for many years, they have often been utilized as the formal background to create the quality, usability and user-experience evaluation models. Some pilot studies of these works can be found in the literature. For example, Lee and Kozar evaluated e-commerce websites with the AHP method [38]. Lin used the AHP method to evaluate online learning courses [39]. Sun and Lin used the fuzzy TOPSIS method for evaluating the competitive advantages of shopping websites [59]. Del Vasto-Terrientes et al. used the ELECTRE-III-H method for the evaluation of travel websites [23]. Bilsel et al. used a hybrid of AHP and Promethee methods for assessing hospital websites [14]. Kaya used an integrated fuzzy AHPTOPSIS methodology for the evaluation of website quality in e-business [33]. Huang et al. used a combination of Simple Additive Weighting, Multiplicative Exponent Weighting, TOPSIS, concordance and conflict investigation techniques to create an e-commerce performance assessment model [29].

In the recent years, a growing popularity of research tools based on eye tracking (ET) can be observed. Although originally they were used mainly in medicine, nowadays we can also find studies on their application in the user experience [20], website quality [18] and usability evaluation [21]. Selected applications of ET in these fields are collected in Table 1. When reviewing the literature, two distinct groups of research emerge. The research in the first group is based on perceptual measurements from ET. The works in the second group are based on a hybrid approach where the ET data is combined with the survey data. The main advantage of using the perceptual data is the fact that the ET provides a set of objective measurements, as opposed to the subjective opinions expressed in the survey data. On the other hand, while the surveys can provide a comprehensive overview of the complete website quality, the ET provides data on a limited group of arbitrarily selected parts of the websites only.

The literature review identifies an interesting research gap – the lack of an integrated approach connecting classical survey methods with empirical data, using dedicated mathematical approach for data aggregation. In practical terms, we propose a comparative analytical study of the eQual classic evaluation method to the crisp and fuzzy variants of the TOPSIS method. A novel approach combining eye tracking and surveys data using mathematical foundations derived from MCDA methodology is presented. The practical authors' contribution is the empirical verification if and how mach the GAZE data influences the overall performance and the rankings of websites' quality, usability and user experience evaluation.

The placement in the e-commerce sector, makes the research important from the method's practical evaluation point of view. Furthermore, focusing on the top e-commerce websites allows the users to remain in the area which does not have a steep learning curve, since the respondents had performed the analyzed activities multiple times before the research begun.

Table 1. Websites evaluation with the usage of eye tracking (ET) and surveys (S)

ET	S	Ref.	Application	Users	Aim of the research	Data analysis methods and results	Criteria
✓	✓	[52]	online shops, online newspapers, company webpages	40	To examine the relation between location typicality and efficiency in finding target web objects on the homepages	Analysis of location typicality, time to first fixation (TFF) and fixations before target (FBT) using Wilcoxon signed-rank tests	2 TFF, FBT
✓	✓	[43]	social commerce	34	Analysis of the impact of the price level and position, and the presentation of the product by a famous person to the fixation time on the website and on the price.	Statistical tests for the time of fixations on the page and price, and gender comparison.	2 fixation time on the page, fixation time on the price time to first
✓	✓	[28]	eTourism 2.0	60	Hypotheses analysis what kind of advertising is more effective.	Statistical analysis, t test. Three separate covariance analyses (ANCOVAs) were computed, with gender, expert level and type of advertisement as independent variables and age as metric covariate.	3 fixation, fixation duration, fixations before
✓	✓	[36]	clinical guidelines on the Web	14	Study of the usefulness of the sites containing medical guidelines for doctors.	Comparison of the task success evaluation to the user experience. Overall performance of the websites was calculated with the geometric mean of the task execution time.	4
✓	-	[57]	e-commerce, B2B	25	Study of the difference in perception of B2B sites by different cultural groups.	Calculation of the correlation, to what extend each of the 7 criteria affect the attractiveness of the pages and comparison of two cultural groups.	7 time to first
✓	-	[68]	online banking	10	Usability study of the electronic banking login interface.	The results consisted of comparison of the numerical data (criteria) obtained during the study and heat maps and AOI trajectory maps. Data obtained during the interview was analyzed.	3 fixation, fixation duration, total time
✓	-	[37]	websites of mobile service providers (telecoms)	44	Comparative evaluation of user experience (UX) and usability.	Basic statistics. Comparison of the obtained results of each criterion for each page (min, max mean, median). For each value: job completion time, time and count of fixations since first click, time to find the target, number of pages viewed during task execution.	3 task completion
✓	-	[7]	e-government websites	9	Study of usefulness of e-government websites.	Basic statistics of the experiment and comparison of the results from the eye tracker with the results from the survey after the experiment.	3 time, fixation duration, fixations count task
✓	-	[48]	online hotel booking websites	16 valid	The purpose of the study was to analyze the impact of images and the size of selection sets on the decision-making process of hotel reservations online.	Based on the data collected in a combined (eye tracking and surveys) experiment, hypotheses were statistically confirmed by comparing the time and number of fixations.	3 completion time, fixation duration, fixations count

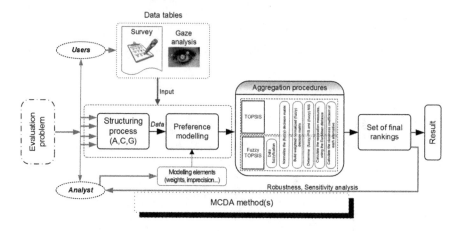

Fig. 1. Proposed evaluation framework.

3   Methodology

The authors' framework is based on the classical survey based approach. However, in addition to the eQual method usage, it was decided to extend the set of input values with the Gaze values for the selected 3-element set of slides for each website. The data was collected using an eye tracker (ET). This approach extends the retrospective (and also prone to the evaluators' subjectivity) survey research with additional empirical data, thus constituting a step towards the objectivity of the website evaluation itself.

It was also decided, that the application of the aggregation technique used inter alia in the eQual method (a derivative of the Simple Additive Weighting method) could be causing an oversimplification of the obtained model. The use of the MCDA mathematical methodology provides the means of comprehensive aggregation of the input data, as well as a more extensive analysis at the aggregation stage or in the final model itself. Apart from the crisp numbers, fuzzy or interval representations of the input data can be used. For example, the application of the American School based MCDA methods provides the ability to obtain an aggregated final ranking including the rating scores, while simultaneously having the possibility to model the preferences, e.g. by altering the weights of the criteria, as well as to analyze the robustness and sensibility of the final model. The scheme of the proposed framework is presented in Fig. 1.

The first element of the framework that needs clarification is the evaluation problem definition. In this paper, the e-commerce websites' usability, quality and user experience evaluation is being evaluated. In the empirical study, we conduct the evaluation of the selected 10 e-commerce websites: Alibaba, Amazon, Apple, BestBuy, eBay, Macy's Rakuten, Staples, Target and Walmart. The choice of the studied websites was based on the analysis of valid rankings of the top e-commerce websites, which were presented inter alia in [1–3] and [4].

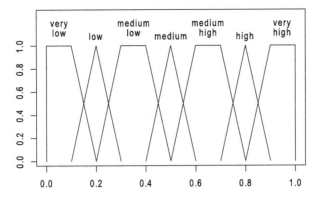

Fig. 2. Trapezoidal Fuzzy Number (TFN) representation of the linguistic weights of the criteria.

The next element of the framework is the stage of the model structuring process. It was assumed here, that the final form of the set of the criteria should contain the full set of criteria C1–C22 contained in the eQual method, along with a set of 6 perceptual criteria:

- E1 – viewers – number of people who have visited the configured areas of interest (AOI);
- E2 – first view [s] – time elapsed in seconds before the area was noticed for the first time;
- E3 – watched time [s] – time spent on a given AOI, expressed in seconds;
- E4 – watched time [%] – time spent on a given AOI, expressed in percent;
- E5 – revisitors – the number of participants who returned to the AOI;
- E6 – revisits – number of revisits to the AOI.

The preference modeling step was repeated multiple times (see the feedback robustness/sensitivity analyzes on Figs. 1, 5 and in Sect. 4.6). Initially, an even distribution of weights was assumed. The detailed weight values in the subsequent steps are presented successively in the results section.

According to Roy [16,53] or Guittoni [27], the choice of the correct aggregation technique (in the form of a particular MCDA method) is an important task. A defined decision problem can be classified into one of representative decision problematics: choice problematic (α), sorting problematic (β) or ranking problematic (γ). Using the guidelines of the authors mentioned above, it was decided that the method should be included in the ranking problematic, retaining the numerical values of the final rating. In addition, the selected MCDA technique should accommodate the possible weights of the criteria while taking into account the different input data forms. Furthermore, a synthetic measurement of the alternatives should be produced by the method. The analysis of the available solutions in the area defined above, indicates the possibility of using the TOPSIS method, which meets all of the aforementioned restrictions. Additionally, some fuzzy developments of the method are available for this aggregation technique. What is more, the TOPSIS method provides a very useful feature to obtain a so-called ideal solution (PIS, positive ideal solution) – which may be an interesting guideline for building successful web interfaces.

The TOPSIS method's algorithm consists of 6 stages [55]. In the first stage, the decision maker choices m alternatives and n criteria which will be utilized for the decision problem analysis. The decision matrix $D[x_{ij}]$ is constructed, where each row represents the decision attributes of each of the alternatives, whereas each column represents a single criterion. Therefore, the x_{ij} element represents a decision attribute of the ith alternative regarding the jth criterion (5):

$$D[x_{ij}] = \begin{pmatrix} x_{11} & x_{12} & x_{13} & \ldots & x_{1n} \\ x_{21} & x_{22} & x_{23} & \ldots & x_{2n} \\ x_{31} & x_{32} & x_{33} & \ldots & x_{3n} \\ \ldots & \ldots & \ldots & \ldots & \ldots \\ x_{m1} & x_{m2} & x_{m3} & \ldots & x_{mn} \end{pmatrix} \tag{5}$$

In the second step of the procedure, the decision matrix is normalized. The decision attributes are normalized separately for each criterion. The benefit and cost criteria are normalized with the formulae from the Eqs. (6) and (7) respectively [44]:

$$r_{ij} = \frac{x_{ij} - min_i(x_{ij})}{max_i(x_{ij}) - min_i(x_{ij})} \tag{6}$$

$$r_{ij} = \frac{max_i(x_{ij}) - x_{ij}}{max_i(x_{ij}) - min_i(x_{ij})} \tag{7}$$

In the third step of the technique, a weighted normalized decision matrix is created. Each element of the matrix is created with the formula (8):

$$v_{ij} = w_j \cdot r_{ij} \tag{8}$$

In the fourth step, the Positive Ideal Solution (PIS) V_j^+ and Negative Ideal Solution (NIS) V_j^- are obtained (9), (10):

$$V_j^+ = \{v_1^+, v_2^+, v_3^+, \ldots, v_n^+\} \tag{9}$$

$$V_j^- = \{v_1^-, v_2^-, v_3^-, \ldots, v_n^-\} \tag{10}$$

In the penultimate step, the Euclidean distances between the i-th alternative and the positive and negative ideal solutions are calculated (11), (12)

$$D_i^+ = \sqrt{\sum_{j=1}^{n}(v_{ij} - v_j^+)^2} \tag{11}$$

$$D_i^- = \sqrt{\sum_{j=1}^{n}(v_{ij} - v_j^-)^2} \tag{12}$$

Eventually, the relative closeness to the ideal solution is calculated using the formula from the Eq. (13) and, based on the obtained closeness coefficient (CCi), the ranking of the alternatives is created.

$$CC_i = \frac{D_i^-}{D_i^- + D_i^+} \tag{13}$$

Nevertheless, as it is noted by the authors of [62], under many conditions, the crisp data is inadequate to model real-life situations, because human judgments on preferences are often vague and difficult to estimate with an exact numerical value. Therefore, they present a fuzzy variant of the TOPSIS method, where the linguistic assessment is used instead of the numerical values. In this variant of the method, the trapezoidal fuzzy numbers $\tilde{n} = (n_1, n_2, n_3, n_4)$ are used as the partial decision attributes and criteria weight instead of crisp numbers. The membership function $\mu_{\tilde{n}}(x)$ for TFN is defined as (14) [32]:

$$\mu_{\tilde{n}}(x) = \begin{cases} 0, & x < n_1, \\ \frac{x-n_1}{n_2-n_1}, & n_1 \leq x \leq n_2, \\ 1, & n_2 \leq x \leq n_3, \\ \frac{x-n_4}{n_4-n_4}, & n_3 \leq x \leq n_4, \\ 0, & x > n_4, \end{cases} \tag{14}$$

The distance between two fuzzy numbers $\tilde{m} = (m_1, m_2, m_3, m_4)$ and $\tilde{n} = (n_1, n_2, n_3, n_4)$ is calculated by using the vertex method (15) [19]:

$$d_v(\tilde{m}, \tilde{n}) = \sqrt{\frac{1}{4}[(m_1 - n_1)^2 + (m_2 - n_2)^2 + (m_2 - n_2)^2 + (m_2 - n_2)^2]} \tag{15}$$

In the authors' framework, the crisp TOPSIS results are compared to the results from the fuzzy variant of the method. The decision attribute for each alternative and each criterion was described by a TFN $\tilde{n} = (n_1, n_2, n_3, n_4)$, based on the input data. The n_1 and n_4 values were set to the minimum and maximum values of the aggregated data for each criterion and each alternative. The n_2 and n_3 values were represented respectively by the difference and the sum of the mean value and the standard deviation of the aggregated data. The weights of importance of the criteria were represented in the research by seven linguistic variables: very low, low, medium low, medium, medium high, high, very high. These variables were also represented in a TFN form and are depicted on Fig. 2.

4   Results

4.1   Empirical Research

An experiment was performed to evaluate the quality, usability and user experience of 10 selected e-commerce websites. The research was based on the survey data from [63] and on the eye tracker (ET) measurements data collected

Table 2. Rank and Evaluation Quality Indexes of the analysed websites [63]

Website	Alibaba A1	Amazon A2	Apple A3	BestBuy A4	eBay A5	Macys A6	Rakuten A7	Staples A8	Target A9	Walmart A10
Rank	4	2	1	7	3	6	9	8	10	5
EQI	68.64%	77.27%	79.20%	66.79%	75.53%	67.42%	64.84%	66.46%	64.41%	68.15%

during an experiment. Originally, the survey results from [63] were based on questionnaires collected from 41 computer-literate users, who evaluated the 10 e-commerce websites. The results were subject to consistency reliability analysis which confirmed their high level of reliability. Subsequently, the classic eQual method was used to calculate the Evaluation Quality Index and thus produce a ranking of the studied websites. The ranking originally obtained in [63] is presented in Table 2. The ET results were obtained from a group of 20 students who, with usage of GazePoint software, were presented slides of the studied websites' home, product and payment pages. The participants were asked to locate a specific element on each displayed page, a so-called area of interest (AOI). Each page was displayed to the participants in a form of slides for the duration of 10 s, with 3 s break between each slide. The eye tracking device collected the data sets E1–E6, which were discussed in Sect. 3. The authors prepared a set of R scripts that facilitated the research presented below and potentially would allow to use the presented approach in practical settings.

4.2 Crisp TOPSIS Approach

In the first step of the research, the averaged values from [63] were used to create a decision matrix for the TOPSIS method. The preference direction for the 22 eQual criteria was set to maximum. An equal weight of 0.5 was assigned to each of the criteria. The obtained decision matrix is presented in Table 3. The closeness coefficient (CCi) of each alternative, along with the obtained ranking is presented in Table 4a.

The comparison of the rankings based on the classic eQual method (EQI) and the crisp TOPSIS method with the eQual criteria (CE) is depicted on Fig. 3a. The horizontal axis of the chart represents the alternative's rank in the EQI ranking, whereas the vertical axis represents the rank obtained in the CE ranking. The analysis of the figure demonstrates that apart from the positions 7 and 8, where a swap of the A4 and A8 alternatives can be observed, the remaining parts of the obtained rankings are consistent. The alternatives A3, A2 and A5, i.e. Apple, Amazon and eBay, are leading in both rankings, whilst the alternatives A7 and A9, i.e. Rakuten and Target, are on the least favorable positions.

4.3 Crisp TOPSIS with Survey and Perceptual Criteria

In the second step, the E1–E6 perceptual criteria were combined with the C1–C22 survey criteria and processed by the crisp TOPSIS method (CEG).

Table 3. Decision matrix for the crisp TOPSIS method based on the eQual criteria

Website	Alibaba A1	Amazon A2	Apple A3	BestBuy A4	eBay A5	Macys A6	Rakuten A7	Staples A8	Target A9	Walmart A10	Weight
C1	4.902	5.610	5.683	5.000	6.024	5.073	4.976	4.927	4.854	5.049	0.5
C2	4.951	5.707	5.415	4.878	5.951	5.000	5.098	4.927	4.756	5.220	0.5
C3	5.000	5.317	5.610	5.000	5.610	4.902	4.805	4.829	4.683	4.829	0.5
C4	4.829	5.390	5.585	4.878	5.634	5.098	4.854	4.659	4.854	5.244	0.5
C5	4.829	5.024	5.976	4.341	4.683	4.707	4.268	4.512	4.220	4.927	0.5
C6	5.098	5.488	6.024	4.561	5.341	5.073	4.707	4.927	4.707	4.805	0.5
C7	4.829	5.366	5.829	4.537	4.878	4.732	4.439	4.732	4.415	4.805	0.5
C8	4.634	5.146	5.415	4.049	4.512	4.585	4.024	4.220	3.683	4.268	0.5
C9	5.000	5.537	5.049	5.073	5.634	4.780	4.805	4.780	4.756	4.537	0.5
C10	4.902	5.537	5.902	5.098	5.683	4.927	5.024	4.805	4.902	4.805	0.5
C11	5.585	5.268	5.488	5.122	5.415	5.512	5.488	5.146	5.561	5.317	0.5
C12	4.951	5.463	5.341	5.268	5.537	4.902	4.732	4.854	5.049	4.610	0.5
C13	4.732	5.537	5.561	5.244	5.512	4.902	4.756	4.707	4.902	4.976	0.5
C14	4.854	5.488	5.171	5.098	5.220	4.634	4.659	4.854	5.024	4.488	0.5
C15	4.927	5.390	5.293	4.854	5.488	4.732	4.512	4.829	4.756	4.951	0.5
C16	4.927	5.829	5.927	4.244	5.878	4.537	4.415	4.488	4.195	4.927	0.5
C17	4.732	5.805	6.000	4.537	5.659	4.537	4.293	4.927	4.317	4.951	0.5
C18	4.732	5.610	5.805	4.707	5.561	4.659	4.390	4.780	4.220	4.902	0.5
C19	3.951	4.927	4.878	3.537	4.049	4.049	3.659	3.756	3.366	3.951	0.5
C20	3.878	4.683	4.293	3.366	3.488	3.537	3.463	3.610	3.146	3.756	0.5
C21	4.780	5.268	5.561	4.829	5.293	4.659	4.268	4.390	4.610	4.732	0.5
C22	4.683	5.610	6.171	4.634	5.268	4.780	4.220	4.683	4.220	4.902	0.5

The weights of each criteria were set, similarly as in Sect. 4.2, to an equally distributed value of 0.5. All criteria's preference directions were set to maximum, except the criterion E2, for which the preference direction was set to minimum. The obtained decision matrix is presented in Table 5 and the ranking along with the alternatives' closeness coefficients (CCi) resulting from the crisp TOPSIS method are presented in Table 4b. The comparison of the CEG ranking with the CE one is presented on Fig. 6a.

It can be observed, that the introduction of the 6 new criteria had a significant influence on the resulting ranking. The alternatives A5 and A10 are the only ones to remain on their positions (3 and 5 respectively). There was a swap between the alternatives on the positions 1 and 2, as well as 9 and 10. The alternative A6 fell from position 6 to 7, and the alternative A8 which was previously on position 7, was demoted to position 8. The most significant change is visible for the alternative A4, which was promoted from position 8 to position 4, which was probably caused by the fact that the alternative A4 achieved very advantageous results in the eye tracker experiment.

4.4 Fuzzy TOPSIS Approach

The results generated by the crisp TOPSIS method are based on the mean values of the criteria. In order to verify if the more advanced knowledge of the individual survey responses and individual eye tracking experiment measurements can influence the final rankings, another evaluation of the websites was performed with the fuzzy variant of the TOPSIS method. The decision matrix constructed from the survey responses is presented in the appendix in Table 9. The value of each criterion is represented by a trapezoidal fuzzy number $\tilde{n} = (n_1, n_2, n_3, n_4)$. A *medium* trapezoidal fuzzy number was assigned as a weight of each of the criteria. The preference direction was set to maximum. The resulting ranking (FE) and the closeness coefficients are presented in Table 4c.

The comparison of the rankings based on the classic EQI and FE rankings is depicted on Fig. 3b. The analysis of the Fig. 6b allows to notice significant differences between the rankings generated by the TOPSIS method in its crisp and fuzzy variants. The positions of the alternatives A5 and A1 were the only ones to remain unchanged. The position of the alternatives A2, A3, A6 and A7 changed by one slot and the positions of the remaining alternatives changed by two slots. It is important to note, however, that despite a slight change in their order, the alternatives A2, A3 and A5 remain the leading websites in both rankings.

Table 4. Ranking of websites based on crisp (a, b) and fuzzy (c, d) TOPSIS methods, and eQual (a, c) or eQual + Eye Tracker (b, d) criteria

		Alibaba A1	Amazon A2	Apple A3	BestBuy A4	eBay A5	Macys A6	Rakuten A7	Staples A8	Target A9	Walmart A10
a	rank	4	2	1	8	3	6	9	7	10	5
	CCi	0.356	0.770	0.850	0.239	0.601	0.301	0.146	0.244	0.137	0.339
b	rank	6	1	2	4	3	7	10	8	9	5
	CCi	0.426	0.774	0.732	0.471	0.723	0.415	0.262	0.407	0.285	0.438
c	rank	4	1	2	6	3	5	10	9	8	7
	CCi	0.522	0.566	0.562	0.512	0.553	0.517	0.500	0.502	0.505	0.508
d	rank	4	3	2	6	1	10	7	9	8	5
	CCi	0.489	0.505	0.540	0.483	0.541	0.480	0.482	0.481	0.481	0.484

4.5 Fuzzy TOPSIS with eQual and Perceptual Criteria

In the fourth step of the research, the perceptual criteria E1–E6 were combined with the survey criteria C1–C22. The E1–E6 criteria were expressed as trapezoidal fuzzy numbers and are presented in Table 6. Each of the new criteria were assigned the *medium* fuzzy weight. The preference direction of the E2 criterion was marked as minimum, and the remaining criteria's direction was marked as maximum. The fuzzy TOPSIS method for the combined criteria produced the ranking (FEG) and scores presented in Table 4d.

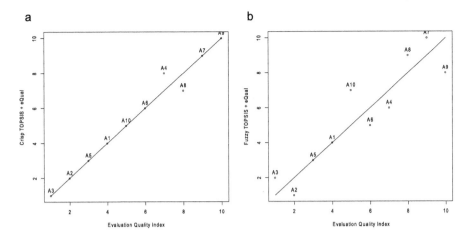

Fig. 3. Comparisons of the rankings generated by the classic eQual and the (a) crisp; (b) fuzzy TOPSIS methods.

When the results of the two fuzzy approaches are compared (see Fig. 6e), it can be observed that the positions of five alternatives, i.e. A1, A3, A4, A8 and A9, remained unchanged. Additionally, although their order changed slightly, the alternatives A2, A3 and A5 are still the leading ones in both rankings. It is worth noting, that the A6 alternative, which dropped only by one slot when the crisp variant was analyzed (see Fig. 6a), fell to the last position of the ranking in the fuzzy variant when the perceptual measurements were taken into account. This fact demonstrates, that the application of fuzzy criteria values instead of the mean values can bring more information to the evaluation procedure.

The analysis of the Fig. 6d, which compares the crisp and fuzzy results based on the combined set of eQual and perceptual criteria, points out that the alternatives can be organized into three groups. The first group, consisting of the alternatives A2, A3 and A5, represents the leading websites, the second group, consisting the alternatives A1, A4 and A10, represents the sites positioned in the middle of the ranking, whereas the third group, consisting of the alternatives A6, A7, A8 and A9, receives the least score in the both methods of evaluation. Eventually, the comparison of the FE and CEG ranking results (see Table 6f) also upholds the alternatives A2, A3 and A5 as the leading ones.

4.6   Sensitivity Analysis

In the next step of the conducted research, the effect that the change of the weight of each criterion has on the final ranking was analyzed. For the crisp versions of the TOPSIS method, the weights of all criteria were set to 50 and then the weight of the analyzed criterion was modified in the range from 1 to 99. In the case of the fuzzy versions of the TOPSIS method, initially all criteria weights were set to *medium* and, subsequently, the weight of the analyzed criterion was

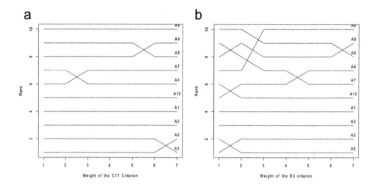

Fig. 4. Sensitivity analysis of the rankings: (a) FEG for the C17 criterion; (b) FEG for the E3 criterion

Table 5. Decision matrix for the crisp TOPSIS method based on the eQual + Gaze criteria

Website	Alibaba A1	Amazon A2	Apple A3	BestBuy A4	eBay A5	Macys A6	Rakuten A7	Staples A8	Target A9	Walmart A10	Weight
C1	4.902	5.610	5.683	5.000	6.024	5.073	4.976	4.927	4.854	5.049	0.5
C2	4.951	5.707	5.415	4.878	5.951	5.000	5.098	4.927	4.756	5.220	0.5
C3	5.000	5.317	5.610	5.000	5.610	4.902	4.805	4.829	4.683	4.829	0.5
C4	4.829	5.390	5.585	4.878	5.634	5.098	4.854	4.659	4.854	5.244	0.5
C5	4.829	5.024	5.976	4.341	4.683	4.707	4.268	4.512	4.220	4.927	0.5
C6	5.098	5.488	6.024	4.561	5.341	5.073	4.707	4.927	4.707	4.805	0.5
C7	4.829	5.366	5.829	4.537	4.878	4.732	4.439	4.732	4.415	4.805	0.5
C8	4.634	5.146	5.415	4.049	4.512	4.585	4.024	4.220	3.683	4.268	0.5
C9	5.000	5.537	5.049	5.073	5.634	4.780	4.805	4.780	4.756	4.537	0.5
C10	4.902	5.537	5.902	5.098	5.683	4.927	5.024	4.805	4.902	4.805	0.5
C11	5.585	5.268	5.488	5.122	5.415	5.512	5.488	5.146	5.561	5.317	0.5
C12	4.951	5.463	5.341	5.268	5.537	4.902	4.732	4.854	5.049	4.610	0.5
C13	4.732	5.537	5.561	5.244	5.512	4.902	4.756	4.707	4.902	4.976	0.5
C14	4.854	5.488	5.171	5.098	5.220	4.634	4.659	4.854	5.024	4.488	0.5
C15	4.927	5.390	5.293	4.854	5.488	4.732	4.512	4.829	4.756	4.951	0.5
C16	4.927	5.829	5.927	4.244	5.878	4.537	4.415	4.488	4.195	4.927	0.5
C17	4.732	5.805	6.000	4.537	5.659	4.537	4.293	4.927	4.317	4.951	0.5
C18	4.732	5.610	5.805	4.707	5.561	4.659	4.390	4.780	4.220	4.902	0.5
C19	3.951	4.927	4.878	3.537	4.049	4.049	3.659	3.756	3.366	3.951	0.5
C20	3.878	4.683	4.293	3.366	3.488	3.537	3.463	3.610	3.146	3.756	0.5
C21	4.780	5.268	5.561	4.829	5.293	4.659	4.268	4.390	4.610	4.732	0.5
C22	4.683	5.610	6.171	4.634	5.268	4.780	4.220	4.683	4.220	4.902	0.5
E1	0.833	0.917	0.883	0.967	0.917	0.783	0.783	0.900	0.883	0.833	0.5
E2	2.316	2.037	2.616	2.022	1.777	2.004	2.437	1.524	2.414	2.207	0.5
E3	1.070	2.080	2.148	2.213	1.955	1.172	1.389	2.772	1.074	0.947	0.5
E4	10.701	20.259	20.619	22.128	19.549	11.712	13.888	27.657	10.739	9.472	0.5
E5	0.667	0.867	0.683	0.900	0.833	0.650	0.733	0.817	0.783	0.667	0.5
E6	2.017	3.717	3.533	3.717	3.400	3.500	3.367	3.350	4.083	2.483	0.5

modified in the range from *very low* to *very high*. The resulting change of the alternatives' ranks was then plotted.

The crisp TOPSIS rankings proved to be very stable. In the ranking based on the eQual data, the first leading alternatives A3, A2 and A5 remain on positions 1, 2 and 3 respectively, regardless of the changes in any of the criteria (see Fig. 5a). The change of the weight of the criteria C1, C2, C5, C7, C11 and C16 weight causes no change of the order of any of the alternatives in the ranking (Fig. 5b). In the case of the ranking based on the combined survey and perceptual criteria (CEG), the alternative A2 has the highest rank and it changes only if the weight of the criterion C11 is increased to over 85 (Fig. 5c).

The ranking based on the survey criteria data processed by the fuzzy variant of the TOPSIS method (FE) is even more stable. Only the change of the weight of the criterion C5 to *very high* causes a swap between the two leading alternatives A2 and A3 (Fig. 5d). No further changes of the ranking can be observed for any change of the weights of the other criteria.

In the case of the ranking obtained from the fuzzy variant of the TOPSIS method based on the combined C1–C22 and E1–E6 criteria (FEG), however, a decrease of stability can be observed. While for the top five alternatives in the ranking only some minor corrections can be observed when the weight of the criteria is decreased to *low* or increased to *high* (Fig. 4a), the changes that can be noticed between the remaining alternatives are more substantial (Fig. 4b). This might be caused by the fact that the values of the perceptual criteria aggregated as trapezoidal fuzzy numbers convey more information than their averaged alternatives in the crisp TOPSIS approach.

4.7 Correlation Analysis

In the last step of the research, the correlation coefficients between the positions in the rankings (Table 7) and the closeness coefficients (CCi, Table 8) obtained by the individual alternatives was analyzed. The values of the correlations confirm the high similarity of the obtained rankings to the one produced by the classic eQual method, with the CE ranking being the most and the FEG ranking being the least similar to the original EQI one. The FEG ranking has considerably

Table 6. Perceptual criteria fragment of the decision matrix for the fuzzy TOPSIS method based on the eQual + Gaze criteria

Criterion E1	E2	E3	E4	E5	E6
A1	(1,0.65,1.02,1) (0.95,1.18,3.46,7.26)	(0.03,0.44,1.7,6.77)	(0.32,4.4,17,67.68)	(0,0.43,0.9,1)	(0,0.93,3.11,10)
A2	(0,0.78,1.06,0) (0,0.85,3.22,0)	(0,1.14,3.02,0)	(0,10.8,29.72,0)	(0,0.7,1.04,0)	(0,2.06,5.38,0)
A3	(1,0.72,1.05,1) (0,1.42,3.81,1.8)	(2.03,1.09,3.21,6.26)	(20.32,9.99,31.25,62.56)	(1,0.45,0.92,1)	(1,1.75,5.32,9)
A4	(1,0.88,1.06,1) (0.53,1.14,2.91,7.1)	(0.05,1.33,3.1,4.02)	(0.48,13.25,31,40.16)	(0,0.75,1.05,1)	(0,2.43,5,10)
A5	(1,0.78,1.06,1) (0,0.49,3.06,3.43)	(2.42,1.01,2.9,8.61)	(24.16,10.1,29,86.08)	(1,0.65,1.02,1)	(1,1.97,4.83,12)
A6	(0,0.58,0.99,1) (0,0.92,3.09,6.44)	(0,0.34,2.01,2.62)	(0,3.35.20.07,26.24)	(0,0.41,0.89,1)	(0,1.49.5.51,9)
A7	(1,0.58,0.99,1) (0,1.24,3.64,7.87)	(0.56,0.64,2.14,6.54)	(5.6,6.42,21.35,65.44)	(0,0.51,0.96,1)	(0,1.84,4.9,8)
A8	(0,0.75,1.05,1) (0,0.36,2.69,6.83)	(0,1.59,3.95,3.2)	(0,15.85,39.46,32)	(0,0.62,1.01,1)	(0,2.01,4.69,15)
A9	(0,0.72,1.05,1) (0,1.38,3.45,9.08)	(0,0.47,1.67,5.74)	(0,4.74,16.73,57.44)	(0,0.58,0.99,1)	(0,2.35,5.82,10)
A10	(0,0.65,1.02,1) (0,1.14,3.27,2.87)	(0,0.28,1.62,7.15)	(0,2.78,16.16,71.52)	(0,0.43,0.9,1)	(0,1.15,3.82,9)

Table 7. Correlations between the positions of the websites in the rankings produced by EQI, crisp TOPSIS with eQual criteria (CE), crisp TOPSIS with eQual+Gaze criteria (CEG), fuzzy TOPSIS with eQual criteria (FE) and fuzzy TOPSIS with eQual+Gaze criteria (FEG).

	EQI	CE	CEG	FE	FEG
EQI	1	0.9879	0.8909	0.9152	0.8061
CE	0.9879	1	0.8424	0.8788	0.7697
CEG	0.8909	0.8424	1	0.8909	0.7818
FE	0.9152	0.8788	0.8909	1	0.7212
FEG	0.8061	0.7697	0.7818	0.7212	1

Table 8. Correlations between the closeness coefficients of the websites in the rankings produced by EQI, crisp TOPSIS with eQual criteria (CE), crisp TOPSIS with eQual+Gaze criteria (CEG), fuzzy TOPSIS with eQual criteria (FE) and fuzzy TOPSIS with eQual+Gaze criteria (FEG)

	EQI	CE	CEG	FE	FEG
EQI	1	0.9946	0.9658	0.9780	0.8855
CE	0.9946	1	0.9538	0.9694	0.8404
CEG	0.9658	0.9538	1	0.9556	0.8317
FE	0.9780	0.9694	0.9556	1	0.8542
FEG	0.8855	0.8404	0.8317	0.8542	1

lower values of correlation coefficients compared to other rankings studied in this paper. This is probably caused by the fact that the trapezoidal fuzzy numbers built on the perceptual data carry more information than the mean values used in the crisp TOPSIS method. Furthermore, the fuzzy TOPSIS method takes into account the uncertainty of the perceptual evaluations, which explains the bigger differences between the FE and FEG rankings than it can be observed between the CE and CEG rankings.

5 Conclusions

E-commerce is a very important sector of online business with billion dollars' worth of sales every year. Due to the high availability and high competition of the online stores, it is crucial to evaluate their quality, usability and user experience systematically.

The authors' contribution in this paper was to create a unique integrated approach connecting classical survey data evaluation methods with empirical data using a dedicated mathematical approach with MCDA methodology foundations. A comparative analytical study of the eQual classic evaluation method to the crisp and fuzzy variants of the TOPSIS method was conducted. A verification that the introduction of the objective gaze measurements data to the subjective questionnaire data impacts the final rankings was performed. A set of six perceptual evaluation criteria was added to the original set of twenty two eQual method criteria.

The problem of the websites' quality, usability and user experience was described in the paper. Selected classic evaluation methods were listed and shortly characterized, and the basics of the eQual method were outlined. Subsequently, the usefulness of the MCDA methods in the data aggregation was discussed and selected works in this area were reviewed. Eventually, an experiment was performed to evaluate a group of 10 selected e-commerce websites with the usage of the presented evaluation framework. The rankings obtained were compared with each other as well as with the ranking generated by the eQual method. Afterwards, the sensitivity analysis was performed on each of the rankings. Ultimately, the correlation analysis of the obtained solutions was performed.

The methodological contribution presented above included the highlights listed below:

- We propose a unique approach that aggregates the data obtained from surveys and gaze measurements.
- We present the introduction of the data aggregation techniques based on the MCDA methodologies.
- By following the multi-criteria approach, we suggest extending the evaluation model with the robustness and sensitivity analyses in order to improve the practical application of the presented approach.
- We verify the possibility of successful usage of the MCDA-based methods in the problem of websites' quality, usability and user experience.

During the research, possible areas of improvement and future work directions were identified. This study was based on the gaze data collected from three arbitrarily selected areas of the analyzed websites. The number of the areas probed could be increased to provide a broader view of the online stores. Additionally, the measurements from all the areas were combined during the study. An alternative approach could be investigated, where the data from each area is used to form its own group of perceptual evaluation criteria. Furthermore, the positive ideal solution obtained during the TOPSIS analyses could be used to create a new set of guidelines for the e-commerce websites' designers. Last, but not least, the presented approach could be generalized and applied to evaluate sites other than online stores, for example e-government or e-banking.

Appendix

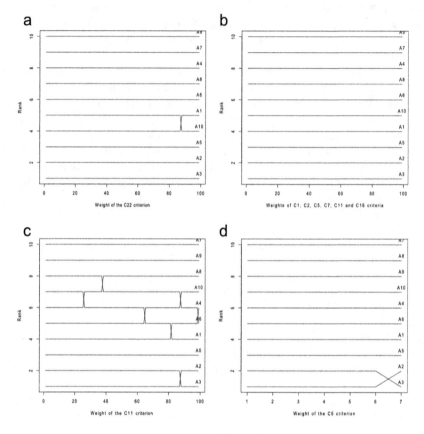

Fig. 5. Sensitivity analysis of the rankings: (a) CE for the C22 criterion (b) CE for the C1, C2, C5, C7, C11 and C16 criteria; (c) CEG for the C11 criterion; (d) FE for the C5 criterion

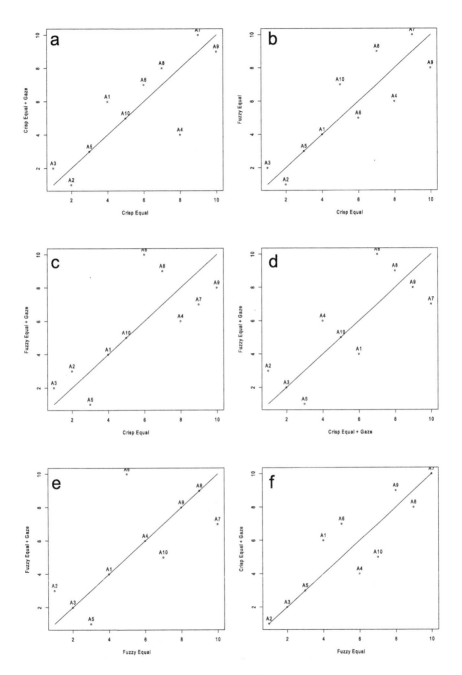

Fig. 6. Comparisons of the rankings.

Table 9. Decision matrix for the fuzzy TOPSIS method based on the eQual criteria

Website	Alibaba A1	Amazon A2	Apple A3	BestBuy A4	eBay A5	Macys A6	Rakuten A7	Staples A8	Target A9	Walmart A10
C1	(2,4.3,5,5,7)	(3,5.01,6.21,7)	(2,5.02,6.35,7)	(2,4.34,5.66,7)	(4,5.56,6.49,7)	(2,4.37,5.77,7)	(1,4.4,5.55,7)	(1,4.22,5.64,7)	(2,4.03,5.68,7)	(1,4.32,5.78,7)
C2	(3,4.34,5.56,7)	(3,5.12,6.29,7)	(2,4.69,6.14,7)	(1,4.18,5.58,7)	(3,5.48,6.42,7)	(2,4.32,5.68,7)	(1,4.44,5.76,7)	(1,4.28,5.57,7)	(2,3.94,5.57,7)	(1,4.47,5.97,7)
C3	(2,4.42,5.58,7)	(2,4.62,6.01,7)	(2,4.91,6.31,7)	(2,4.28,5.72,7)	(2,5.02,6.2,7)	(1,4.14,5.67,7)	(1,4.1,5.51,7)	(1,4.02,5.64,7)	(1,3.81,5.56,7)	(1,4.18,5.48,7)
C4	(2,4.23,5.43,7)	(3,4.74,6.04,7)	(2,4.9,6.27,7)	(2,4.21,5.54,7)	(2,5.05,6.22,7)	(2,4.42,5.78,7)	(2,4.2,5.51,7)	(1,3.93,5.39,7)	(2,4.04,5.66,7)	(1,4.64,5.84,7)
C5	(1,4.11,5.54,7)	(1,4.21,5.83,7)	(3,5.32,6.63,7)	(1,3.5,5.18,7)	(1,4.01,5.36,7)	(2,3.98,5.43,7)	(1,3.57,4.97,7)	(1,3.84,5.18,7)	(1,3.31,5.13,7)	(1,4.23,5.63,7)
C6	(2,4.53,5.67,7)	(2,4.89,6.08,7)	(3,5.45,6.6,7)	(2,3.84,5.28,7)	(2,4.78,5.91,7)	(2,4.44,5.71,7)	(2,4.06,5.35,7)	(1,4.17,5.69,7)	(2,3.97,5.44,7)	(1,4.11,5.5,7)
C7	(3,4.2,5.46,7)	(3,4.82,5.91,7)	(2,5.13,6.53,7)	(1,3.79,5.28,7)	(1,4.15,5.6,7)	(2,4.02,5.45,7)	(1,3.88,5,7)	(1,4.02,5.44,7)	(1,3.59,5.24,7)	(1,4.2,5.41,7)
C8	(1,3.85,5.41,7)	(2,4.5,5.79,7)	(1,4.56,6.27,7)	(1,3.27,4.82,7)	(1,3.82,5.2,7)	(1,3.73,5.44,7)	(1,3.13,4.92,7)	(1,3.46,4.98,7)	(1,2.79,4.57,6)	(1,3.49,5.04,7)
C9	(2,4.23,5.77,7)	(3,4.98,6.1,7)	(1,4.19,5.91,7)	(2,4.38,5.77,7)	(3,5.06,6.21,7)	(2,4.03,5.53,7)	(2,4.14,5.47,7)	(1,4.01,5.55,7)	(2,4.15,5.37,7)	(1,3.86,5.22,7)
C10	(2,4.25,5.55,7)	(3,4.99,6.09,7)	(1,5.19,6.62,7)	(2,4.48,5.72,7)	(3,5.11,6.26,7)	(2,4.23,5.62,7)	(2,4.45,5.6,7)	(1,4.12,5.49,7)	(2,4.31,5.49,7)	(1,4.13,5.48,7)
C11	(2,4.92,6.25,7)	(4,4.73,5.8,7)	(4,4.97,6,7)	(4,4.65,5.6,7)	(3,4.84,5.98,7)	(3,4.87,6.16,7)	(1,4.75,6.23,7)	(1,4.41,5.88,7)	(2,4.93,6.19,7)	(2,4.64,5.99,7)
C12	(2,4.26,5.64,7)	(3,4.9,6.02,7)	(2,4.6,6.08,7)	(2,4.62,5.92,7)	(2,5,6.07,7)	(2,4.27,5.53,7)	(1,4.05,5.41,7)	(1,4.14,5.57,7)	(2,4.37,5.73,7)	(1,3.89,5.32,7)
C13	(2,4.12,5.34,7)	(3,5.04,6.04,7)	(3,4.93,6.19,7)	(1,4.53,5.96,7)	(2,4.92,6.11,7)	(2,4.24,5.56,7)	(1,4.12,5.4,7)	(1,3.98,5.43,7)	(2,4.25,5.55,7)	(1,4.27,5.68,7)
C14	(2,4.17,5.54,7)	(3,4.91,6.07,7)	(1,4.33,6.01,7)	(2,4.38,5.85,7)	(2,4.58,5.86,7)	(1,3.95,5.32,7)	(1,4.03,5.29,7)	(1,4.11,5.6,7)	(3,4.37,5.68,7)	(1,3.73,5.25,7)
C15	(2,4.25,5.6,7)	(3,4.81,5.97,7)	(1,4.59,5.99,7)	(1,4.08,5.62,7)	(3,4.86,6.11,7)	(1,3.97,5.49,7)	(2,3.92,5.11,7)	(2,4.12,5.54,7)	(1,3.97,5.54,7)	(1,4.25,5.65,7)
C16	(1,4.27,5.58,7)	(2,5.21,6.45,7)	(2,5.28,6.57,7)	(1,3.58,4.91,7)	(2,5.28,6.47,7)	(1,3.8,5.27,7)	(1,3.78,5.05,7)	(1,3.84,5.13,7)	(2,3.6,4.79,7)	(1,4.21,5.65,7)
C17	(1,3.87,5.59,7)	(3,5.2,6.41,7)	(3,5.42,6.58,7)	(1,3.8,5.27,7)	(2,4.99,6.33,7)	(1,3.69,5.38,7)	(1,3.6,4.98,7)	(1,4.14,5.71,7)	(1,3.56,5.07,7)	(1,4.09,5.82,7)
C18	(1,4.01,5.46,7)	(3,5.02,6.2,7)	(3,5.15,6.46,7)	(2,4.06,5.35,7)	(3,4.99,6.13,7)	(1,3.86,5.45,7)	(1,3.71,5.07,7)	(1,4.09,5.47,7)	(1,3.49,4.95,7)	(1,4.07,5.74,7)
C19	(1,3.21,4.69,7)	(1,4.28,5.57,7)	(1,3.98,5.77,7)	(1,2.72,4.35,6)	(1,3.17,4.93,7)	(1,3.18,4.92,7)	(1,2.93,4.39,7)	(1,2.88,4.63,7)	(1,2.56,4.17,7)	(1,3.25,4.65,7)
C20	(1,3.09,4.66,7)	(1,3.97,5.39,7)	(1,3.29,5.29,7)	(1,2.54,4.19,7)	(1,2.64,4.33,7)	(1,2.75,4.33,7)	(1,2.7,4.22,6)	(1,2.76,4.46,6)	(1,2.28,4.01,7)	(1,2.97,4.54,7)
C21	(1,3.89,5.67,7)	(1,4.53,6.01,7)	(2,4.93,6.19,7)	(1,4.07,5.59,7)	(2,4.68,5.91,7)	(1,3.79,5.53,7)	(1,3.5,5.03,7)	(1,3.54,5.24,7)	(1,3.81,5.41,7)	(1,3.94,5.52,7)
C22	(1,3.86,5.51,7)	(1,4.92,6.3,7)	(2,5.55,6.79,7)	(1,3.82,5.45,7)	(1,4.49,6.05,7)	(1,3.88,5.68,7)	(1,3.44,5,7)	(1,3.94,5.43,7)	(1,3.47,4.97,7)	(1,4.11,5.7,7)

References

1. Information economy report 2015 (2015). http://unctad.org/en/PublicationsLibrary/ier2015_en.pdf
2. Top 10 e-commerce companies in the world 2015 (2015). http://www.mbaskool.com/fun-corner/top-brand-lists/13991-top-10-ecommerce-companies-in-the-world-2015.html?start=1
3. Alexa's top e-commerce websites (2016). http://www.alexa.com/topsites/category;3/Business/E-Commerce
4. World's top 10 ecommerce sites (2016). http://www.dollarfry.com/worlds-top-10-ecommerce-sites-alexa-rankbasis/
5. Ahn, T., Ryu, S., Han, I.: The impact of the online and offline features on the user acceptance of internet shopping malls. Electron. Commerce Res. Appl. **3**(4), 405–420 (2005)
6. Akinci, S., Atilgan-Inan, E., Aksoy, S.: Re-assessment of ES-qual and E-RecS-qual in a pure service setting. J. Bus. Res. **63**(3), 232–240 (2010)
7. Albayrak, D., Cagiltay, K.: Analyzing Turkish e-government websites by eye tracking. In: 2013 Joint Conference of the 23rd International Workshop on Software Measurement and the 2013 Eighth International Conference on Software Process and Product Measurement (IWSM-MENSURA), pp. 225–230. IEEE (2013)
8. Barnes, S., Vidgen, R.: Webqual: an exploration of website quality. In: ECIS 2000 Proceedings, p. 74 (2000)
9. Barnes, S.J., Vidgen, R.: An evaluation of cyber-bookshops: the webqual method. Int. J. Electron. Commerce **6**(1), 11–30 (2001)
10. Barnes, S.J., Vidgen, R.: Measuring web site quality improvements: a case study of the forum on strategic management knowledge exchange. Ind. Manage. Data Syst. **103**(5), 297–309 (2003)
11. Barnes, S.J., Vidgen, R.: The equal approach to the assessment of e-commerce quality: a longitudinal study of internet bookstories (2005)
12. Barnes, S.J., Vidgen, R.T.: Data triangulation in action: using comment analysis to refine web quality metrics. In: ECIS 2005 Proceedings, p. 24 (2005)
13. Barnes, S.J., Vidgen, R.T.: Data triangulation and web quality metrics: a case study in e-government. Inf. Manage. **43**(6), 767–777 (2006)
14. Bilsel, R.U., Büyüközkan, G., Ruan, D.: A fuzzy preference-ranking model for a quality evaluation of hospital web sites. Int. J. Intell. Syst. **21**(11), 1181–1197 (2006)
15. Boulding, W., Kalra, A., Staelin, R., Zeithaml, V.A.: A dynamic process model of service quality: from expectations to behavioral intentions. J. Market. Res. **30**(1), 7 (1993)
16. Bouyssou, D., Jacquet-Lagrèze, E., Perny, P., Slowiński, R., Vanderpooten, D., Vincke, P.: Aiding Decisions with Multiple Criteria: Essays in Honor of Bernard Roy. International Series in Operations Research & Management Science, vol. 44. Springer, New York (2002). doi:10.1007/978-1-4615-0843-4
17. Brans, J.P., Mareschal, B.: Promethee methods. In: Figueira, J., Greco, S., Ehrogott, M. (eds.) Multiple Criteria Decision Analysis: State of the Art Surveys. International Series in Operations Research & Management Science, vol. 78. Springer, New York (2005). doi:10.1007/0-387-23081-5_5
18. Chang Lee, K., Wook Chae, S.: Exploring the effect of the human brand on consumers' decision quality in online shopping: an eye-tracking approach. Online Inf. Rev. **37**(1), 83–100 (2013)

19. Chen, C.T.: Extensions of the topsis for group decision-making under fuzzy environment. Fuzzy Sets Syst. **114**(1), 1–9 (2000)
20. Chen, L., Pu, P.: Eye-tracking study of user behavior in recommender interfaces. In: Bra, P., Kobsa, A., Chin, D. (eds.) UMAP 2010. LNCS, vol. 6075, pp. 375–380. Springer, Heidelberg (2010). doi:10.1007/978-3-642-13470-8_35
21. Cowen, L., Ball, L.J., Delin, J.: An eye movement analysis of web page usability. In: Faulkner, X., Finlay, J., Détienne, F. (eds.) People and Computers XVI - Memorable Yet Invisible. Springer, London (2002). doi:10.1007/978-1-4471-0105-5_19
22. Cunliffe, D.: Developing usable web sites-a review and model. Internet Res. **10**(4), 295–308 (2000)
23. Del Vasto-Terrientes, L., Valls, A., Slowinski, R., Zielniewicz, P.: Electre-III-H: an outranking-based decision aiding method for hierarchically structured criteria. Expert Syst. Appl. **42**(11), 4910–4926 (2015)
24. Elling, S., Lentz, L., Jong, M.: Website evaluation questionnaire: development of a research-based tool for evaluating informational websites. In: Wimmer, M.A., Scholl, J., Grönlund, Å. (eds.) EGOV 2007. LNCS, vol. 4656, pp. 293–304. Springer, Heidelberg (2007). doi:10.1007/978-3-540-74444-3_25
25. Grigera, J., Garrido, A., Panach, J.I., Distante, D., Rossi, G.: Assessing refactorings for usability in e-commerce applications. Empir. Softw. Eng. **21**(3), 1224–1271 (2016). http://dx.doi.org/10.1007/s10664-015-9384-6
26. Guitouni, A., Martel, J.M.: Tentative guidelines to help choosing an appropriate MCDA method. Eur. J. Oper. Res. **109**(2), 501–521 (1998)
27. Guitouni, A., Martel, J.M., Vincke, P., North, P.: A framework to choose a discrete multicriterion aggregation procedure. Defence Research Establishment Valcatier (DREV) (1998)
28. Hernández-Méndez, J., Muñoz-Leiva, F.: What type of online advertising is most effective for etourism 2.0? An eye tracking study based on the characteristics of tourists. Comput. Hum. Behav. **50**, 618–625 (2015)
29. Huang, J., Jiang, X., Tang, Q.: An e-commerce performance assessment model: its development and an initial test on e-commerce applications in the retail sector of China. Inf. Manage. **46**(2), 100–108 (2009)
30. Hwang, C.L., Lai, Y.J., Liu, T.Y.: A new approach for multiple objective decision making. Comput. Oper. Res. **20**(8), 889–899 (1993)
31. Kabir, G., Hasin, M.: Comparative analysis of topsis and fuzzy topsis for the evaluation of travel website service quality. Int. J. Qual. Res. **6**(3), 169–185 (2012)
32. Kauffman, A., Gupta, M.M.: Introduction to Fuzzy Arithmetic, Theory and Application (1991)
33. Kaya, T.: Multi-attribute evaluation of website quality in e-business using an integrated fuzzy ahptopsis methodology. Int. J. Comput. Intell. Syst. **3**(3), 301–314 (2010)
34. Kemp, S.: Digital in 2016 (2016). https://www.slideshare.net/wearesocialsg/digital-in-2016/537
35. Kemp, S.: Digital in 2017 global overview (2017). https://www.slideshare.netwearesocialsgdigital-in-2017-global-overview
36. Khodambashi, S., Gilstad, H., Nytrø, Ø.: Usability evaluation of clinical guidelines on the web using eye-tracker. Stud. Health Technol. Inform. **228**, 95 (2016)
37. Kruger, R.M., Gelderblom, H., Beukes, W.: The value of comparative usability and UX evaluation for e-commerce organisations (2016)
38. Lee, Y., Kozar, K.A.: Investigating the effect of website quality on e-business success: an analytic hierarchy process (AHP) approach. Decis. Support Syst. **42**(3), 1383–1401 (2006)

39. Lin, H.F.: An application of fuzzy AHP for evaluating course website quality. Comput. Educ. **54**(4), 877–888 (2010)
40. Lindner, M.: Online sales will reach $523 billion by 2020 in the U.S., January 2016. https://www.digitalcommerce360.com/2016/01/29/online-sales-will-reach-523-billion-2020-us/
41. Matell, M.S., Jacoby, J.: Is there an optimal number of alternatives for likert scale items? Study i: reliability and validity. Educ. Psychol. Measure. **31**(3), 657–674 (1971)
42. McSpadden, K.: You now have a shorter attention span than a goldfish. Time Online Magaz. (2015). Accessed 7 May 2016. http://time.com/3858309/attention-spans-goldfish/
43. Menon, R.V., Sigurdsson, V., Larsen, N.M., Fagerstrøm, A., Foxall, G.R.: Consumer attention to price in social commerce: eye tracking patterns in retail clothing. J. Bus. Res. **69**(11), 5008–5013 (2016)
44. Milani, A., Shanian, A., Madoliat, R., Nemes, J.: The effect of normalization norms in multiple attribute decision making models: a case study in gear material selection. Struct. Multidisc. Optim. **29**(4), 312–318 (2005)
45. Min, H.: International supplier selection: a multi-attribute utility approach. Int. J. Phys. Distrib. Logist. Manage. **24**(5), 24–33 (1994)
46. Nielsen, J.: Usability Engineering. Elsevier, Amsterdam (1994)
47. Olsson, M.: Build a Profitable Online Business: The No-Nonsense Guide, 1st edn. Apress, Berkely (2013)
48. Pan, B., Zhang, L.: An eyetracking study on online hotel decision making: the effects of images and number of options (2016)
49. Parasuraman, A., Zeithaml, V.A., Malhotra, A.: Es-qual a multiple-item scale for assessing electronic service quality. J. Serv. Res. **7**(3), 213–233 (2005)
50. Parasuraman, A., Zeithaml, V.A., Berry, L.L.: Servqual: a multiple-item scale for measuring consumer perc. J. Retail. **64**(1), 12 (1988)
51. Paul, R.: Just how big is the ecommerce market? You will never guess!, June 2015. http://blog.lemonstand.com/just-how-big-is-the-ecommerce-market-youll-never-guess/
52. Roth, S.P., Tuch, A.N., Mekler, E.D., Bargas-Avila, J.A., Opwis, K.: Location matters, especially for non-salient features-an eye-tracking study on the effects of web object placement on different types of websites. Int. J. Hum. Comput. Stud. **71**(3), 228–235 (2013)
53. Roy, B.: Multicriteria Methodology for Decision Aiding, pp. 51–68. Springer Science & Business Media, November 11, 2013
54. Seixedo, C., Tereso, A.P.: A multicriteria decision aid software application for selecting MCDA software using AHP. In: 2nd International Conference on Engineering Optimization (EngOpt2010) (2010)
55. Shih, H.S., Shyur, H.J., Lee, E.S.: An extension of topsis for group decision making. Math. Comput. Model. **45**(7), 801–813 (2007)
56. Sohaib, O., Kang, K.: Assessing web content accessibility of e-commerce websites for people with disabilities (2016)
57. Štrach, P., Slivkin, N.: Adaptation needed: eye-tracking study of cross-cultural differences in perception of b2b websites (2017)
58. Strzelecki, A., Furmankiewicz, M., Ziuziański, P.: The use of management dashboard in monitoring the efficiency of the internet advertising campaigns illustrated on the example of google analytics. Studia Ekonomiczne **296**, 136–150 (2016)
59. Sun, C.C., Lin, G.T.: Using fuzzy topsis method for evaluating the competitive advantages of shopping websites. Expert Syst. Appl. **36**(9), 11764–11771 (2009)

60. Treu, S.: Need for multi-aspect measures to support evaluation of complex human-computer interfaces. In: Fourth Annual Symposium on Human Interaction with Complex Systems, Proceedings, pp. 182–191. IEEE (1998)
61. Twidale, M.B., Nichols, D.M.: Exploring usability discussions in open source development. In: Proceedings of the 38th Annual Hawaii International Conference on System Sciences (HICSS 2005), p. 198c. IEEE (2005)
62. Wang, J.W., Cheng, C.H., Huang, K.C.: Fuzzy hierarchical topsis for supplier selection. Appl. Soft Comput. **9**(1), 377–386 (2009)
63. Wątróbski, J., Ziemba, P., Jankowski, J., Wolski, W.: Pequal-e-commerce websites quality evaluation methodology. In: 2016 Federated Conference on Computer Science and Information Systems (FedCSIS), pp. 1317–1327. IEEE (2016)
64. Weatherhead, R.: Say it quick, say it well-the attention span of a modern internet consumer. Guardian Online **19** (2012). https://www.theguardian.com/media-network/media-network-blog/2012/mar/19/attention-span-internet-consumer
65. Webb, H., Webb, L.: Business to consumer electronic commerce website quality: integrating information and service dimensions. In: AMCIS 2001 Proceedings, p. 111 (2001)
66. Webb, H.W., Webb, L.A.: Sitequal: an integrated measure of web site quality. J. Enterpr. Inf. Manage. **17**(6), 430–440 (2004)
67. Yang, Z., Cai, S., Zhou, Z., Zhou, N.: Development and validation of an instrument to measure user perceived service quality of information presenting web portals. Inf. Manage. **42**(4), 575–589 (2005)
68. Yuan, X., Guo, M., Ren, F., Peng, F.: Usability analysis of online bank login interface based on eye tracking experiment. Sensors Transducers **165**(2), 203 (2014)

Model Supporting Social Media Hiring

Lucie Bohmova$^{(\boxtimes)}$, Antonin Pavlicek **(iD)** , and Petr Doucek

University of Economics, Prague, Czech Republic
{Lucie.Bohmova,Antonin.Pavlicek,Petr.Doucek}@vse.cz

Abstract. Creation of model which supports Social media hiring. Analysis based on analysing existing frames for employees hiring with help of social media. Finding a gap and fill it with own model is key for this method. For creation of own model for social media hiring has been used application which extracted user data from Facebook. As predictors for users' behaviour on the social media networks have been used personality test MBTI. With the help of Cluster Analysis and machine learning (decision trees) have been created a Stochastic predictive model, that determine a personality type. The Sample of N = 960 Facebook users, that joined hiring app follows that users on the social media network such as Facebook share a lot of useful information for hiring. Verification of this model on the test sample confirmed correctness of prediction personalities category MBTI in a range of 68% to 84% and by individual cases with confidentiality in a range of 43% to 81%. Model for social media hiring contains a guide/manual for automatic data mining of users, particularly from social media network Facebook. It also contains a suggestion how to analyse mined data.

Keywords: Model · Social networking sites · Cluster analysis · Data mining

1 Introduction

Organizations have a problem with finding right employees. Traditional methods do not work properly anymore thanks to the low unemployment rate, high labor demand, decreasing number of economically active people [7] and characteristic features of newly incoming generations (such as Y and Z)[1] to the labor market or newly established trend of shared economics.

A partial solution can be exploiting of social networking sites (SNS) - innovative and low-cost solution, the prospective source of very high-quality specialist or just ordinary workers across all generations. [7] However for companies is very difficult to find out which SNS they should use for hiring and how to take advantage of their potential. [7, 8] Authors offer as a solution a model for social media hiring that is introduced in this article.

1.1 Current Models/Frames

Although the term "model" can have many definitions, for our purpose we can understand it as simplified form of visualization of researched reality frame. Model is

---

[1] Modern, independent, lacking sense of work commitments with leisure time as a priority.

© Springer International Publishing AG 2017
S. Wrycza and J. Maślankowski (Eds.): SIGSAND/PLAIS 2017, LNBIP 300, pp. 81–95, 2017.
DOI: 10.1007/978-3-319-66996-0_6

built according to particular rules that allow imitating a behavior and properties of visualized reality [13].

From the analysis of existing frames follows, that only a few of them work on social media hiring. Down below are described chosen frames for hiring with a focus on that parts that includes Social Media Networks.

Company Oracle together with Institute Human Capital released a methodology (Social Recruiting Guide) that includes seven questions organization should answer before they start social media hiring campaign [19]. Once the organization answers these questions, they should start to work on the strategy plan of hiring. The company needs to find where it has followers and opponents and then they can start to take advantage of their strategy. Certainly, the part of the strategy must be metrics for results measuring.

Online and Social Media Recruitment Model according to Ladkin and Buhalis [10] classify social media networks into the hiring process and in the scope of six questions lead the organization to the selection of the most suitable social media networks. This model is designed especially for hospitality, nevertheless is usable even for other industries.

Hongzhi [20] proposed a model TCAM (Temporal Context-Aware Mixture) that deals with the analysis of user's behavior on the social media networks and it catch their preferences. This model revealed information that conduct of users on the social media networks is determined by personal interest and time context. Another model from the same authors is DTCAM (Dynamic Temporal Context-Aware Mixture Model) that loosely follows up a previous model and combines an influence of both factors on modeling users conduct. The interest of users are not always stable and can evolve in time we expand TCAM model to model of Dynamic Temporal Context-Aware Mixture, that catch up with changing interests of users. Models TCAM and DTCAM are not primary for hiring on social media networks. However, both fit in thanks to the analysis of user's conduct.

Another examples are Proposed Practical Model for Social Media Driven Collaborator Recruitment [9], Model of strategic planning and exploiting of Social Media Networks [17], Methodology ICIMS [11], Model COBRA [14], Social Media Activity Model [2] etc.

1.2   Summary of Existing Methods

Current methods are very wide and general and they do not deal with particular guide for every social media network itself. Amongst existed methods, that deal with social media hiring can be mentioned ICIMS or methods from companies Oracle and Jobvite. From the analysis freely accessible hiring methods follows that only few of them is about social media hiring.

Models of social media hiring should be based on strategies, organization structures, company processes and workflows etc. Models that deals with Hiring processes is so many and organization can choose even according to single steps of the process. Generally, the hiring process is summarized by HRM strategy models that show the entire process from different perspectives. Examples can be Model of agreement,

Harvard system, Model of the best practices, Contingency model. However, models that really deal with Social media networks hiring are lacking. [1]

The weak spot of current models is "not sufficient usage of social media networks" as a tool that is able to provide additional information, reference about potential candidate or even acquiring new employee.

Hence Authors decided to fill these gaps in proposed model that will serve to organization as a support in hiring and this model will also predict a personality type based on their Social Media Networks behaviour.

2   Solution

The main goal of this work is to create and verify the usefulness of the model for social media hiring that will be valid and useable for every kind of organization and job position. Therefore is very important to set up criteria for model creation.

2.1   Data Extraction App from Social Media Network Facebook

The key starting point for creating the model is an own app named "Prace na Miru" developed by authors. This App mines data about users and it runs at web page www. prace-na-miru.eu. Data that are gathered are stored into database MongoDB as JSON. Page and Database are hosted via heroku.com. A programming language is JavaScript. Code has been written in framework Node.js. Limitation of application is that you can´t update data of users according to the current activity. Workflow of extraction is in Fig. 1. Candidate goes to the website of "Prace na Miru", where he can find a login button to Facebook. After inserting his log in credentials there is initiated a window where he can find what is going to be downloaded. Candidate gives a permission to download a data that will be downloaded into the database.

Fig. 1. The data extraction workflow (Source: Authors)

Information about application "Práce na míru" has been spread via the email newsletter to the target audience that is students and fresh alumni of the University of Economics in Prague. Also, the application has been promoted on social medias in particular groups reaching target group and in case study during work related event named "Šance 2016" and on the website of university career center. 960 unique applicants have signed on to the application in period October 2016 to January 2017.

Analysis of data that were downloaded automatically via Prace na Miru
Thanks to tools Excel and KNIME authors moved these data into the readable form and cleared them a little bit. From the analysis, we can see interesting results.

Table 1. A number of users in % that have public accessible information according to the categories on Facebook profile (Source: Authors)

Categories of publicly available data	Nr. of users in %	Categories of publicly available data	Nr. of users in %	Categories of publicly available data	Nr. of users in %
Gender	99	Tagged places	76	Favourite books	35
Login devices	92	Favourite music	73	Languages	30
Friendlists	91	Actual location	73	Relationship details	30
User_birthday	89	Hometown	67	Games activity	29
Profile photo	87	Favourite TV series	66	Quote	16
Likes	84	Favourite films	56	Interest in a particular person	16
Education	83	Friend's posts on Timeline	56	Favourite sports	14
Email	82	Gallery of Photos	53	Bio	12
Events	81	Favourite Athletes	50	Religion	10
Own timeline posts	81	Favourite Athlete's Teams	45	Favourite inspirational person	8
Videos	77	Relationship	44	Politics	7
Photos	76	Work	36	Website	6

Table 1 is showing percentage representation of public accessible information about users according the particular category. Users share information such as gender, device (used for log-in), the list of friends, date of birth, etc. Some of this information are suitable for recruiting. On the other hand, users do not share information about religion, politics opinion, inspiring people. Authors are mentioning only interesting or important conclusions related to the recruitment of employees.

From the analysis can be seen the basic demography information about users such as gender, age, and education. Representation of women and men is almost equal. Age representation by birth year is shown below, see Graf 1. The biggest group of users according to the date of birth is amongst years 1986 to 1997 that is in right fit with the target group. Primary it is a generation Y and rarely generation Z (Graph 1).

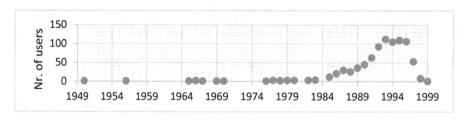

Graph 1. Users date of birth by years (Source: Authors)

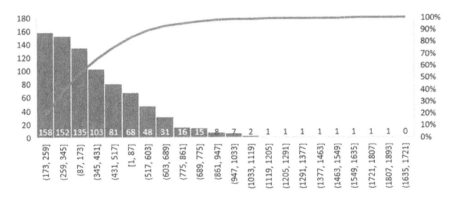

Graph 2. Number of friends – Pater's graph (Source: Authors)

Authors have also been interested in a number of friends of users, see Graph 2. These numbers are rapidly represented at the scale of 87 to 259 friends. The range is 1 853 that tells us there are significant differences. Therefore, we are more interested in median instead of the average. The median is 288 friends.

Profile photography have as public accessible information 87% of users, which means from the hiring perspective an option to verify a candidate also on other social media networks. Users usually have in average Eighteen photography as public accessible information in the gallery.

Regarding analysis, the authors have also been focused on correlation concerning published data in particular categories. They have used Spearman's correlation coefficient, see Table 2. According to the results of the Table 2, there has been proven an existence of low bond tightness amongst the amount of public accessible data for categories: favorite book and series (0.32), favorite game and book (0.32), favorite game and movie (0.31), friends and event (0.31). Medium-bond tightness has been found out in categories: like and event (0.47), favorite movie and book (0.52), favorite movie and music (0.49). Significant-bond tightness has been found out in the

Table 2. Correlation of public accessible information of users amongst categories (Source: Authors)

Categories of publicly available data	Friendlists	TV series	Athletes	Books	Events	Games	Music	Likes	Films
Friendlists	1	0.09	0.11	0.09	**0.31**	-0.06	0.16	0.29	0.04
TV series	0.09	1	0.17	**0.32**	0.11	0.23	**0.57**	0.09	**0.57**
Athletes	0.11	0.17	1	0.15	0.07	0.25	0.16	0.03	0.12
Books	0.09	**0.32**	0.15	1	-0.09	**0.32**	0.27	0.07	**0.52**
Events	**0.31**	0.11	0.07	-0.09	1	-0.01	0.14	**0.47**	0.03
Games	-0.06	0.23	0.25	**0.32**	-0.01	1	0.11	0.04	**0.31**
Music	0.16	**0.57**	0.16	0.27	0.14	0.11	1	0.22	**0.49**
Likes	0.29	0.09	0.03	0.07	**0.47**	0.04	0.22	1	0.12
Films	0.04	**0.57**	0.12	**0.52**	0.03	**0.31**	**0.49**	0.12	1

Fig. 2. Word cloud (Source: Authors)

categories: favorite music and series (0.57) and favorite music and movie (0.57). It follows that users sharing information about favorite series also share information about favorite music and movies.

Post on the timeline is a public accessible information in 56% of user cases. These posts contain in average 73 signs, which means short messages. Median is 39 signs. Word cloud has shown us words that are repeated in the posts of users on their Facebook wall, more in Fig. 2. English words are used very often as well as Czech text without diacritics. The largest category is www, that means that users are sharing links to the external websites.

Czech Republic is dominating in the count of visited places that users have visited (5 742) following by USA (648), Germany (341) and Italy (265). Only 36% of users have published information about Employer as public. The most often we can see AIESEC (2%), Deloitte (1%) and Accenture (1%). Other Employers are mentioned only in the order of units. One of the main reason can be that users are mostly students without work related experience. Mostly users like pages that are related to University of Economics, official UoE Facebook page, faculties Facebook pages and student organization Facebook pages (AIESEC, Student Union, Club of Investors). In average users on Facebook liked 24 pages. In the category, favourite series users prefer mostly comedies. In average users published six favourite series. The most favourite movies are Harry Potter, Forrest Gump, and Avatar. In average users published nine movies. Books that users like are Harry Potter, "Konec prokrastinace", and "Deníček moderního fotra". In average, they have on profiles four favourite books. Very interesting is differentiate music styles. Users listen preferably Linking Park, Tomas Klus, and The Beatles. Average of 16 items they have on their profiles.

2.2 Predictors for Evaluation of User's Behaviour on the Social Media Network

Next starting point for model construction are predictors for evaluation of user's behaviour on the social media network. In terms of hiring are the best predictors that goes directly out of the personalities test. Therefore, was necessary to specify the requirements at a suitable test in dependency of model purpose that are:

- evaluation of personal characteristics,
- evaluation of interpersonal characteristics,
- evaluation of work characteristics,
- relevancy for hiring,
- speed,
- transparency,
- option to fill the test online from everywhere,
- immediate evaluation without other expenses (Psychologist).

Requirements stated above met MBTI personality test [12], that has been used in the test. MBTI test determines personality types of potential candidates. In Practice is usually used in Human Resources, where it is used while creating job positions and recruiting. It is a part of psychological tests [17]. MBTI test determines personality type of potential candidates that is based on combination of four basic characteristics groups:

- perception of surrounding environment,
- way of obtaining information,
- way of evaluating information,
- life style.

Every criterion mentioned above offers Dichotomius assessment, see Table 3.

MBTI test and its evaluation

To all users that have been logged in to "Prace na Míru" were sent the MBTI test. Response rate was at this point 50.4%. Results of this TEST helped to determine personality types and characteristic features of questioned users and compare them with their behaviour on the social media networks. The sample was 484 unique people.

In view of the distribution of individual group criteria is the ratio between the extrovert and the introvert users is quite balanced along with thinking and feeling. There is a big difference between senses and intuition, along with judgment and perception.

Graph 3 shows the number of users by year of birth and by group criteria. It can be seen from the graph, the year of birth does not affect the distribution of group criteria, so we can approach both, graduates and students in the same way, while creating PM model.

The results of the MBTI test are compared according to the criteria of the Facebook user behaviour predictors while using the Pajek and BigML software tools. Pajek serves to find clusters that are used as predictors, and BigML has the task of creating a decision tree and its graphical representation.

2.3   Legal Boundaries

A significant limitation for the organization while gathering and verifying information on Facebook about the candidate is that the organization must keep in mind the Act about Personal data protection (in CZ is Act no. 101/2000 Sb.) and Labour Legislation. At the same time, Organization can not commit discrimination behaviour (in CZ is Act no. 435/2004 Sb.).

Table 3. Categories of MBTI personality types and their characteristics (Source: [18])

Criteria	Sign	Group	Pros/cons attribute	Characteristic
Perception of surrounding environment	E	Extroversion	+	Focused on people's world, many friends, talk too much, kindly
			−	Shaking off matters that need focus
	I	Introversion	+	Focused on inner world of thoughts, thinking about stuff, good listeners, reserved
			−	Afraid of social contact
Way of obtaining information	S	Sences	+	Focused on present, measureable results, facts, do not trust own intuition, do not search for problems, working continuously, realistic
			−	Issues while looking into future, formulation of conceptions
	N	Intuition	+	Focus on the future or the past are more theoretical than practical; Often thinking about more things at the same time, preferring general answers, failing to follow the procedures, having a developed fantasy
			−	Neglect of details.
Way of evaluating information	T	thinking	+	Logical decision making, better remember numbers and pictures than names and faces, impersonal, focus on performance, prefer the truth before agreement
			−	Problems with workplace relationships
	F	Feeling	+	Decision-making based on feelings and values, interest in others, prefer harmony to truth, empathy
			−	For them It is worse to solve impersonal problems
Life style	J	Reasoning	+	They like to plan, organize, do things thoroughly, on time (closed issues)
			−	Shaking off the rest
	P	Perception	+	Flexible, spontaneous, do not care, like unknown (open problems)
			−	They postpone binding decisions and planning

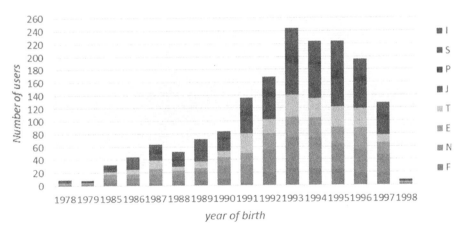

Graph 3. Number of users according to year of birth and MBTI criteria (Source: Authors)

3   Creation of Model for Social Media Hiring

The purpose of the Employee Recruitment Model is implementation into a problematic context, which is primarily the recruitment of suitable candidates for RPC VŠE services (Career centre of the University of Economics) and internship programs within the xPORT Business Business Accelerator in University of Economics. The target group of users are students and fresh graduates of University of Economics.

The design of the model is based on the CRISP-DM methodology, which consists of six basic phases.

(1) *Phase of understanding the problem*
The phase of understanding the problem was carried out in several steps, while defining the research problem and the main aim of this work, the theoretical part of the thesis and the research carried out by the authors [5].

(2) *Phase of understanding the data*
The data comprehension phase is based on "Analysis of data downloaded automatized using the Prace na Miru application" and the determination of appropriate parameters for assessing the behaviour of users according to the MBTI Personality Test in the section "MBTI Test and its Evaluation".
Into the model creation are inserted only categories of data that users share as publicly accessible information in 50% and above. More Table 1.

(3) *Phase of data preparation*
The preparation of the data was based on the selected analytical tool Pajek [16], which serves for our cause as software support for cluster analysis.

Cluster Analysis
In this work is used, the hierarchical clustering method, called the Ward method. It is based on analysis of variance. It merges clusters if there is a minimum sum of squares. Object distances are measured with a square Euclidean distance, see Formula 1.

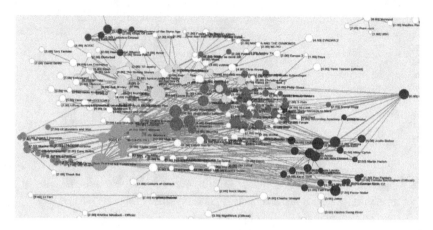

Fig. 3. Clusters according to the connection for attribute favourite TV series (Source: Authors)

$$d_5(u, v) = \sqrt{\sum_{\substack{s = 1 \\ s \neq u, v}}^{n} \left((q_{us} - q_{vs})^2 + (q_{su} - q_{sv})^2 \right) + p \cdot \left((q_{uu} - q_{uv})^2 + (q_{uv} - q_{vu})^2 \right)} \quad (1)$$

For cluster analysis have been used attributes as favourite Music, Favourite TV series, favourite movie and favourite athlete.

Graphical output is net graph, that use colours to highlight created clusters for an attribute favourite TV series, see Fig. 3.

(4) *Modelling phase*

In this phase, we sculpture a decision trees with a help of the tool BigML [3].

Decision trees

For creation of decision trees authors have used the algorithm CART (Classification and Regression Trees) that is generating binary tree. As a criterion for branching is used Gini index, which is one of the function of pollution i(t). Function is defined as:

$$\Delta i(t) = -\sum_{k=1}^{k} P^2(k|t_r) + \sum_{k=1}^{k} P^2(k|t_l) + \sum_{k=1}^{k} P^2(k|t_p). \quad (2)$$

Where p (k | t_r, t_l, t_p) are conditional probabilities, t_r is the parent node, t_l left descendant, tp the right descendant, and k are indexes of the dependent variable class (k = 1,..., K).

For each target area (favourite book, music etc.), the authors created 4 decision trees that determine one of the personality categories, see E or I, S or N, T or F, J or P. In total 16 decision trees.

Figure 4 specifically illustrates the category - obtaining information. The beginning of the tree shows that the key factor is cluster E. The description of the branch is bold grayed out in the figure below: If a user in his profile on Facebook has marked his favourite TV series falling within the E and H clusters and he did not mark one TV series that falls into A, F, D, G, B, then he fits with 90.36% confidence into characteristics (N - intuition) from MBTI personality categories. In this way we can easily read the rest of the branches of the tree.

While using BigML, you can detect data types of individual columns and divide data into separate instances. In the next step, it is possible to use the selected number of

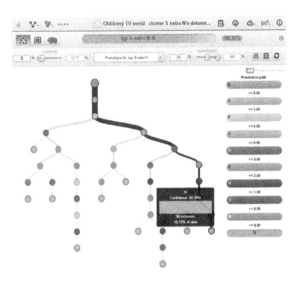

Fig. 4. Decision tree (favourite TV series) for MBTI category - obtaining information (Source: Authors)

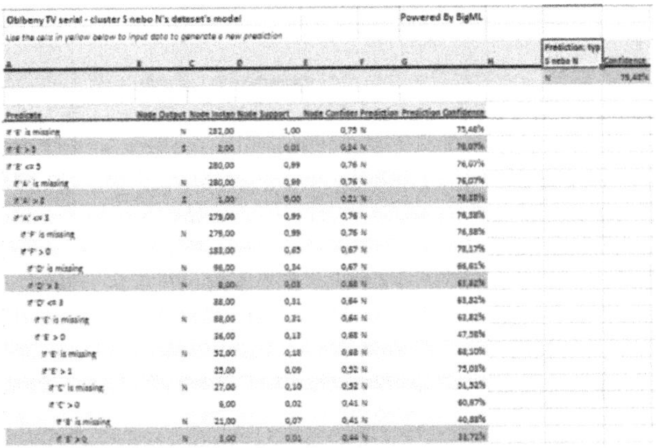

Fig. 5. Form of predictive model for MBTI (Source: Authors)

instances to create a model above which predictions can be made. Figure 5 thus shows the form of the predictive model for determining the MBTI type according to the clusters of particular categories of data (here specifically for the popular TV series), which originated from the machine learning of the decision tree.

As a result of modelling phase is model PM, see Fig. 6.

(5) *Evaluation phase*

Formal verification of the model to support recruitment
On training data (N = 484), the PM model learned the decision about personality types. PM validation took place on new dates collected from users who signed up for the

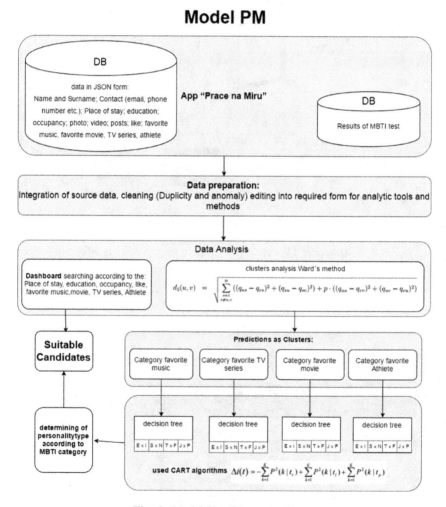

Fig. 6. Model PM (Source: Authors)

Prace na Miru application between February and March 2017. There were 198 people in the target group.

Validation was performed using predictive models that originated from the decision trees for that attribute and MBTI personality category. Thus, the author of the data verification has made publicly available information from Facebook for all four categories (favourite music, a favourite TV series and a favourite athlete).

(6) *Deployment phase*

The final phase is a deployment of the PM model into real use of RPC VŠE and xPORT VŠE Business Accelerator. The environment for which the PM model has been created is constantly changing, and the importance of different data for proper prediction, so the model needs to be continually scanned, expanded and updated to maintain reliability and accuracy.

4 Discussion

Research limitations
A model for social media networks hiring is not suitable for finding and evaluating all people on the labour market, but only those who have a social media network account. The model also does not ensure finding suitable candidates, but it only selects people who are registered in an application and thanks to it, it can extract user data on a given social network. At the same time, the model is affected by the segment of users who log in to the data extraction application.

A necessary condition for selecting social networks that can be used by the model is the openness of the social networking development environment (API) in order to integrate the proposed application for extracting user data.

The Model for social media networks hiring further reflects restrictions based on restrictions legislation, which makes it impossible to use all available social information in the practical application of the model.

Further restrictions can be found in the type of candidates enrolled in the application for extracting user data (Prace na Míru), which are primarily the target group of students and fresh graduates of the University of Economics.

When creating a PM model, the authors are aware of possible model distortions, despite testing the model on real data. Model distortions may be of the false correlation type, the development sequence or the missing middle member [12].

Benefits of the model
- Proving the creation and validation of the Model for Employee Recruitment.
- Filling a gap in existing models for recruiting.
- Satisfying of informational needs of organization while recruiting.
- Prediction of personality type based on behaviour on social networks.

Analysis of existing models to support employee recruitment through social networks, their benefits and limitations.

Possible ideas for further research are

- Create an automated solution for other social networks like LinkedIn and Twitter.
- Adding another students from other Universities into app Prace na Miru.
- Create a comprehensive methodology to support recruitment through social networking.
- Adding a dictionary for emotional colouring of words into the model.

5 Conclusion

Frameworks or models that would give to organizations a clear guidance on the effective use of social media networks for recruitment are lacking. There is no added value in existing frameworks in the form of evaluating user's conduct on the social media networks. Therefore, the authors used the MBTI personality test to help diagnose the characteristics of the candidates, such as the perception of the environment or the way information is obtained and processed, when using a model to support recruitment.

The model includes an automated solution for downloading user data, specifically from the Facebook social network. Also included is a proposal for analytical data processing, specifically the model describes cluster analysis, decision trees and a predictive model to determine the personality type (based on the MBTI test).

Verification of the model on test data confirmed its accuracy of the prediction of the MBTI personality category ranging between 68% to 84% in individual cases with confidentiality ranging between 43% to 81%.

References

1. Armstrong, M.: Řízení lidských zdrojů : nejnovější trendy a postupy. Grada Publishing (2008)
2. Bender, J.L., et al.: Ethics and privacy implications of using the internet and social media to recruit participants for health research: a privacy-by-design framework for online recruitment. J. Med. Internet Res. **19**(4), e104 (2017)
3. BigML. https://bigml.com/
4. Boehmova, L., Malinova, L.: Facebook user's privacy in recruitment process. In: Petr, D., et al. (eds.) IDIM-2013: Information Technology Human Values, Innovation and Economy, pp. 159–166. Verlag Trauner, Praha (2013)
5. ČSÚ: Informační technologie v domácnostech a mezi jednotlivci (2015). https://www.czso.cz/csu/czso/domacnosti_a_jednotlivci
6. ČSÚ: Projekce obyvatelstva České republiky do roku 2100 (2013). https://www.czso.cz/csu/czso/projekce-obyvatelstva-ceske-republiky-do-roku-2100-n-fu4s64b8h4
7. HR news: Twitter pomáhá hledat práci https://www.hrnews.cz/lidske-zdroje/trendy-id-148711/twitter-pomaha-hledat-praci-id-738597
8. Jobvite: Social Recruiting Survey (2014). https://www.jobvite.com/wp-content/uploads/2014/10/Jobvite_SocialRecruiting_Survey2014.pdf

9. Khatri, C., et al.: Social media and internet driven study recruitment: evaluating a new model for promoting collaborator engagement and participation. PLoS ONE **10**(3), e0118899 (2015)
10. Ladkin, A., Buhalis, D.: Online and social media recruitment: Hospitality employer and prospective employee considerations. Int. J. Contemp. Hosp. Manag. **28**(2), 327–345 (2016)
11. Lauby, S.: 4 Steps to Creating a Successful Social Recruiting Strategy.pdf (2016). https://www.icims.com/sites/www.icims.com/files/public/hei_assets/4%20Steps%20to%20Creating%20a%20Successful%20Social%20Recruiting%20Strategy.pdf
12. Mattie, M.: Revisiting understanding entrepreneurs using the myers-briggs type indicator. J. Mark. Dev. Compet. **9**(2), 114–119 (2015)
13. Molnár, Z.: Pokročilé metody vědecké práce. Profess Consulting, Praha (2012)
14. Muntinga, D.G., et al.: Introducing COBRAs. Int. J. Advert. **30**(1), 13–46 (2011)
15. Oracle: Social Recruiting Guide: How to Effectively Use Social Networks (2012) http://www.oracle.com/us/media1/effectively-use-social-networks-1720586.pdf
16. Pajek. http://mrvar.fdv.uni-lj.si/pajek/
17. Pyne, L.: Plan Your Social Media Strategy, https://socialhubsite.com/plan-social-media-strategy/
18. Quenk, N.L.: Essentials of Myers-Briggs Type Indicator Assessment. Wiley, Hoboken (2008)
19. Vajner, L.: Výběr pracovníků do týmu. Grada Publishing a.s. (2007)
20. Yin, H., et al.: Dynamic user modeling in social media systems. ACM Trans. Inf. Syst. **33**(3), 10:1–10:44 (2015)

Information Systems Development

Essential Challenges in Business Systems Modeling

Václav Řepa[✉]

Department of Information Technologies, Faculty of Informatics and Statistics,
University of Economics, Prague, Prague, Czech Republic
repa@vse.cz

Abstract. This paper argues for the need to respect so-called 'Principle of Modeling' and its consequences in the information system development methodologies. Principle of Modeling expresses the presumption that the objective basis for the implementation of the information system in the organization must be constituted by real facts existing outside of and independently of the organization. In other words, information system as an information infrastructure of some business system is always a model of the Real World. The Philosophical Framework for Business System Modeling is used as a platform for the discussion about basic aspects of the Real World and their relationships which should be covered by the information system in terms of the Principle of Modeling. Then some important consequences following from the framework in mutually connected fields of Business Processes Modeling, Conceptual Modeling, and Information System Development are discussed.

Keywords: Information systems development · Principle of modeling · Business process model · Conceptual model · Object orientation · Process orientation

1 Introduction

One of the essential principles of the information systems development methodologies is the Principle of Modeling. Principle of Modeling expresses the presumption that the objective basis for the implementation of the information system in the organization must be constituted by real facts existing outside of and independently of the organization. In other words, information system as an information infrastructure of some business system is always a model of the Real World.

The roots of the Principle of Modeling are closely connected with the technique of Normalization of Data Structures firstly introduced by E.F. Codd in [3] and then elaborated in further detail together with R.F. Boyce in [2]. Although the original Codd's intention was mainly technical and located in the field of database system design, this technique started uncovering the essential Principle of Modeling as a generally valid principle in the field of information systems development. The principle has been later used by Peter Chen who followed-up the Codd's ideas by introducing the 'entity-relationship model' that 'adopts the natural view that the real world consists of entities and relationships and incorporates some of the important semantic information

© Springer International Publishing AG 2017
S. Wrycza and J. Maślankowski (Eds.): SIGSAND/PLAIS 2017, LNBIP 300, pp. 99–110, 2017.
DOI: 10.1007/978-3-319-66996-0_7

about the *real world*' [9]. This way Chen switched from the traditional thinking just about the organization of data to thinking about the real world and its reflection in data later known as the 'conceptual modeling'. During the following decades this way of thinking has been established as the essential general principle which manifests itself not just in the field of data but in all substantial dimensions of information system development.

Principle of Modeling significantly increased the possibilities of ISD methodologies to assure the quality of the information system in the analysis phase of its development. If the contents of the information system is fully determined by the Real World then the quality of the information system should be measurable with the attributes of the Real World. So the ability to achieve the proper quality of the information system is directly related to the ability to uncover the proper attributes of the Real World. The correctness of the information system follows from the truthfulness and completeness of the Real World models used in the process of its development. Based on these ideas the ISD methodologies can be equipped with detailed quality rules following from the attributes of the Real World. These rules can cover both main meanings of the model quality: correctness and completeness. Information system should support its user with correct (i.e. truly) as well as complete information about the relevant part of the Real World.

Despite its four decades long existence, the full respect to the Principle of Modeling and its essential consequences in the field of quality of the contents of the information system is still not usual in contemporary information system development methodologies. Instead of the orientation on objective categories of truthfulness and completeness of the picture of the Real World in the information system most methodologies prefer relative and subsidiary categories for measuring the quality of the system contents like 'user requirements', 'user satisfaction' or even 'user pleasure'.

In this paper we argue for the need to respect the Principle of modeling and its consequences in the information system development methodologies. We introduce the Philosophical Framework for Business System Modeling as a platform for the discussion about basic aspects of the Real World and their relationships which should be covered by the information system in terms of the Principle of Modeling. Using the framework we describe four basic dimensions of the model of the Real World and their natural attributes together with related methods and techniques from the field of information systems analysis. Special attention we pay to the relationships of particular dimensions as important field for the consistency of models.

Finally, we discuss the most important emergent consequences of the nature of basic dimension of the Real World informatics model in the mutually connected fields of Business Processes Modeling, Conceptual Modeling, and Information System Development.

2   Philosophical Framework for Business System Modeling

The main purpose for creating the framework was the need to comprehensively understand all important aspects of the Real World which should be reflected in the information system. This need is based on the idea that information system as an information infrastructure of the business (Real World) system is intended to support the business system with the right, truthful and up-to-date information about its state, as well as its history and possible relevant future events.

For such comprehensive understanding the Real World we need to have the general idea about basic dimensions of the Real World. This idea should be as much as possible general because we aim to use as much as possible knowledge from different research fields focused on the Real World including philosophical disciplines, especially logic. Because of its needful generality we call our framework for business system modeling 'philosophical'.

The framework is based on the following premise: *The picture of the given business domain (Real World) is determined by two basic phenomena:* **being** *and* **behavior** *and two basic views:* **system view** *and* **particular (temporal) view**.

Being represents the Real World as it is (can be, must be, etc.). Real World being covers the basic facts about the existence and possible changes of the Real World objects and their relationships, and can be formally described by means of the modal logic. On the other hand, **behavior** represents the happening in the Real World as a consequence of the acting of the Real World actors in terms of achieving goals, executing plans, etc. (intentional behavior). Such behavior can be formally expressed by means of the process-oriented description. **System view** sees the Real World as a system of particular elements. As the primary purpose is to describe the attributes of the system, such model has to cover the whole system. The system point of view requires the abstraction of individual attributes of system parts which excludes especially modeling of their temporal aspects because they are always partial (see the following paragraph). The system model thus can be also characterized as a static view. *Particular (temporal) view* focuses on Real World events and their consequential changes. To be precise enough such model cannot cover the whole system but just its part. Temporal Real World aspects can be formally modeled by means of the algorithmic description which principally excludes any parallelism. Therefore, each particular model has to be made from the point of view of a single element of the system – temporal view disallows the description of the system characteristics.

By the combination of these two phenomena with these two views we can obtain four essential types of models (see Fig. 1):

(1) The model of the *Real World Modality*, as a static view of being, describes the system of *Real World objects* and their possible mutual relationships.

(2) The model of the *Real World Causality*, as a temporal view of being, describes possible *states in the life of the particular Real World object* and possible transitions among them.

(3) The model of *Collaboration*, as a static view of behavior, describes the *system of business processes* and their mutual relationships. Regarding the necessary intentional character of behavior the relationships among processes always mean their collaboration in order to achieve the defined goals.

(4) The model of *Acting*, as a temporal view of behavior, describes the chains of actions in the particular business process intended to achieve the given process goal under possible circumstances.

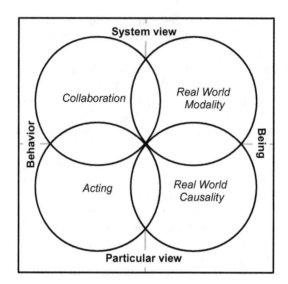

Fig. 1. Philosophical framework for business system modeling (Source: author)

Figure 1 shows not only four basic types of models but also the importance of their mutual intersections which represent the fact that the different types of models are not completely mutually exclusive. Some facts about the Real World are expressed just once in the corresponding type of model. Some other facts are expressed more times, in different models from different points of view. It is simply because basic phenomena and views used in the framework are really not exclusive. Being without behavior does not make sense because each change in the Real World is always a consequence of some action which, seen from the business system perspective, is driven by some purpose. Similarly, particulars do not make a sense without a system which they belong to. Each element of the given particular model is determined by the corresponding system model which defines its necessary context. This awareness of the basic relationships among different informatics models arising from the fact that all models describe the same complex Real World is a base for so-called consistency rules which characterize the high quality methodologies as it is also discussed in the following paragraphs.

Figure 2 describes the kinds of analytical models typically used in the information systems development methodologies which correspond to the particular model types defined in the framework.

For the purpose of this paper the *Real World Modality* means the static Real World rules in terms of the basic (i.e. 'alethic') modal logic. This view of the Real World is in informatics represented by the traditional data-oriented conception of the **conceptual model** represented by [9] for instance. This model describes which Real World objects can (must) be related to which Real World objects at which circumstances.

For the purpose of this paper the *Real World Causality* means the temporal Real World rules in terms of tense logic. This view of the Real World can be in informatics represented by the model of the **object life cycle**. This model describes the causality of

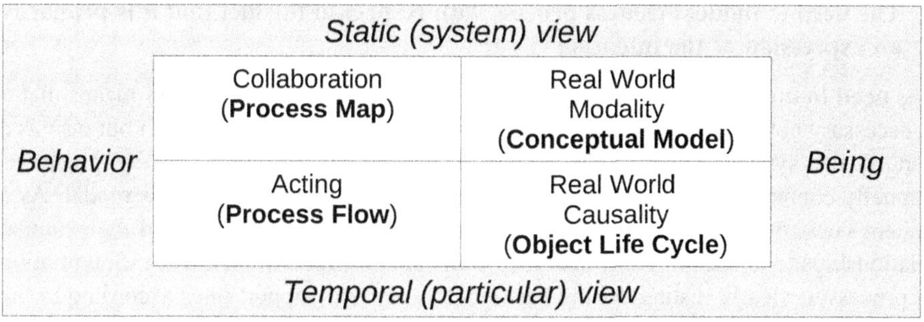

Static (system) view		
Behavior	Collaboration (**Process Map**)	Real World Modality (**Conceptual Model**)	Being
	Acting (**Process Flow**)	Real World Causality (**Object Life Cycle**)	
Temporal (particular) view		

Fig. 2. Kinds of business system models in informatics (Source: author)

the evolution of the Real World object in terms of defined states and their possible sequentiality under specified circumstances.

By the model of *collaboration* we mean the mutual context of business processes: their mutual relationships. Regarding the natural intentionality of behavior in the business system the relationships of the business processes have to be interpreted as their collaboration on achieving the goals. This view of the behavior in the Real World can be in informatics represented by the **Process Map**. The most commonly used notation for the process map comes from the methodology by Erikson and Penker [5].

By the model of *acting* we mean the algorithmic description of one business process which covers all important variants of the behavior of actors in terms of the particular process goal. This view of the behavior in the Real World can be in informatics represented by the **Process Flow Diagram**. Besides the 'de iure' standard Business Process Model & Notation (BPMN) [1] there are more concurrent commonly accepted notations for the process flow diagram like eEPC from ARIS methodology [15], IDEF3 [10] from IDEF, or some of UML extensions, the Eriksson/Penker's one for instance [5].

3   Emergent Consequences

The complex view of the Real World models represented by the framework allows seeing the basic consequences following from the described facts in particular connected fields. In the following paragraphs we discuss those of them which seem to be critically important for the information systems development regarding especially the current state of the art in the fields of IS development methodologies, languages and tools.

In the field of Business Processes Modeling
In the field of Business Processes Modeling the framework uncovers two main actual challenges for current methodologies:

- **The need to model not just the process flow but also the process system**: the global model of processes (usually called Process Map).

- **The need to model business process with respect to the fact that it is primarily an expression of the intention**

The **need to model** not just the process flow but also **the process system** means that it is necessary not only to model the process as a process (i.e. how to run it) but also as a part of the system of processes which is a collection of collaborating processes mutually connected with services. We call this model the Global Process model. As a system view, this model shows the system parts (business processes) and their mutual relationships (cooperation) and that way it allows the needed functional differentiation of processes; clearly distinguishing between the key and support ones according to the business nature of processes expressed in [7]. Unfortunately, this need is still not sufficiently reflected by the current BPM methodologies as it is visible at the state of the art of business process modeling languages. For example BPMN [1], even if it is established as a worldwide standard in the field of business processes modeling, it is still mainly oriented just on the description of internal algorithmic structure of a business process and disregards the global view on the system of mutually cooperating processes. The only way of modeling the cooperation of different processes in BPMN is using 'swimming pools and lanes' in the Collaboration Diagram. Unfortunately, the global aspects of the system of business processes cannot be sufficiently described this way nor its completeness ensured. The BPMN primarily views processes as sequences of actions in the time line. However, the global model requires seeing processes primarily as objects (relatively independent of the time), distinguishing different kinds of them (especially the key versus support ones), describing their global attributes (like the goal, reason, type of customer, etc.), and recognizing their essential relationships to other processes which all is obviously impossible to describe as a process flow.

The above criticized insufficiency of BPMN can be eliminated using the additional model which completes BPMN with needed object-oriented point of view. The most standard way is to use the Eriksson-Penker process diagram [5] as a complement to the BPMN diagram. Eriksson-Penker Notation [5] was created as an extension of Unified Modelling Language (UML) [16] which corresponds with the 'object nature' of the global view on processes discussed above. This notation distinguishes between the 'Business Process View' which illustrates the interaction between different processes and the 'Business Behavioral View' which describes the individual behavior of the actors of one particular process. This way it respects the important difference between the global object-oriented view of a process system and the detailed process-oriented view of a single process. The detailed explanation of the methodical need for global model of processes as well as related criticism of the BPMN can be also found in [14].

Business process details should be modeled with respect to the fact that it is primarily an **expression of the intention**. Intentionality, or more traditionally purposefulness, is also very important topic for the ideas Business Process Management Automation in general, particularly robotics and similar fields. In the legendary article [11], which is usually regarded as the root of cybernetics, the authors expressed the idea which essentially influenced the later development of cybernetics: *'all purposeful behavior may be considered to require negative feed-back'*. The concept of negative feed-back is explained there as follows: *'...the behavior of an object is controlled by the margin of error at which the object stands at a given time with* *reference to a relatively specific goal. The feed-back is then negative, that is, the signals from the goal are used to restrict outputs which would otherwise go beyond the goal'.*

According to the basic work in the field of process-driven management [7], business process always follows some goal. The goal is a fundamental attribute of a business process as it is regularly used in matured methodologies like in [5] for instance. That means that *business process is always an intentional process*, more exactly the process of purposeful behavior of an interested object following some goal. For instance the behavior of the process manager is undoubtedly an intentional behavior which follows the goal of the process.

Taking into the account the definition of purposeful behavior discussed above, it can be said that every business process, as it is an intentional kind of process, have to have some negative feed-back which ensures restriction of its outputs in order to keep them in the margins of its goal. In the case of the business process the feed-back is represented by the input to the process from its environment which is causally connected with some process output. The value of the input should influence the following behavior of the process in terms of keeping it within the margins of its goal. This means that 'intermediate' inputs to the process (i.e. none-starting inputs to the process coming between its starting and end points) are critically important parts of the business process distinguishing it from other, non-intentional (i.e. non-business), processes. When working with processes we have to take into the account even the time dimension; every input to the process from its environment has to be synchronized with the process run. Thus, in each part of the process where some input which influences the following process run is expected the process state has to be placed. The process state means such points in the process structure where nothing can be done before the input to the process occurs, i.e. the point of waiting for the input. Process state thus represents the essential need to synchronize the process run with expected events. This need follows from the fact that the event is always an objective external influence and thus it must be respected. From the physical point of view such respect means synchronization – waiting for the event.

Not all methodologies and process modeling languages respect the concept of a process state. It is pretty well respected in IDEF [10], partially in ARIS [15] and not at all in BPMN [1]. As a 'de iure' standard BPMN do not recognize the concept of a process state there is no other way than to express the process state with the general symbol for synchronization – the 'AND gate'. Some further discussion about process states and other important consequences of the intentionality in business process models, especially regarding the standard BPMN, can be found in [12, 14] and is illustrated with Fig. 3.

In the field of Conceptual Modeling

The general emerging challenge in the field of Conceptual Modeling which follows from the framework is **the need to model also the system dynamics which means in the case of the conceptual model the essential causal rules**. Causality of the evolution of an object can be modeled via so-called object (entity) life cycle. For that purpose the UML is a well prepared language as it allows modeling of life cycles with the State Chart. Moreover, using UML the life cycle of an object class can be modeled

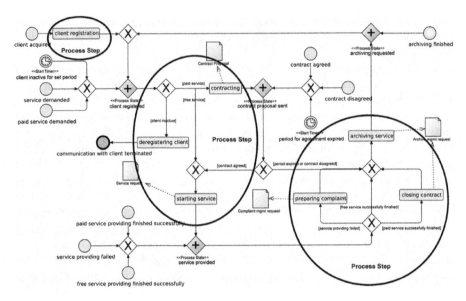

Fig. 3. Example of the use of process states in the BPMN language. (Source: [14])

in the context of other connected classes which is generally described with the Class Diagram. The language specification defines the essential relationships of both diagrams [16]. So in this case the problem does not lie in the language but in the conceptual modeling methodologies which are mostly still anchored in the traditional data-based conception of the conceptual modeling. In this conception the conceptual model is clearly static view of objects and their essential and thus stable relationships. All temporal aspects connected with modeled objects are then typically regarded as a matter of business processes and out of the scope of the conceptual model. Nevertheless, as the Philosophical Framework for Business System Modeling shows, there is a substantial difference between the dynamics of objects (representing the dynamics of being) and the dynamics represented by business processes as intentional chains of actions. Therefore, we believe that life cycles of the conceptual objects can be regarded as an integral part of the conceptual model allowing modeling of not just the static modality of the Real World but also a causal - temporal aspects of the Real World modality in terms of the temporal logic [4].

This way we argue for increasing the scope of conceptual modeling by making the life cycle diagram (State Chart from the UML) a regular diagram for conceptual modeling of object details which should be regarded as a new generation of the theory of conceptual modeling. Requirements for such extension also follows from the 'problem of identity' in connection with the temporal aspects of the Real World and the certain insufficiency of the UML for modeling them identified even in [6, 8].

Some further argumentation for the need to overcome this traditional limitation of the conceptual modeling by object life cycles can be found in [13].

In the field of Information Systems Development

The argumentation from previous paragraphs about the need to model not just static aspects of the Real World but also its dynamics with careful distinguishing between the causal and intentional kind dynamics is critically important also from the information system point of view.

The attempts to 'clear database solutions' are still frequent in the information systems development methodologies. Many researchers and developers still believe that the well designed database built on the precise data analysis (of the Real World) is a sufficient condition for building the whole information system because the functionality of the system is always subordinated to the possibilities of the database. This opinion is supported with the popularity and usefulness of the database tools supporting the standardized system functionality connected with the database in terms of the basic database operations of storing and retrieving data and supported with the integrity definitions which can cover most of the standard functionality of the information system. Another fact supporting such opinion is the approach of the development methodologies to the phenomenon of business processes. The business meaning of processes in terms of business-driven management (represented by [7] for instance) is usually regarded as something out of the scope of the information system development. System developers usually regard business process as just a description of the way of using the information system which can be expressed as a simple use-case.

The problem is that such approach ignores the intentional character of the behavior of business actors. Not only data requirement but also behavior of business actors have to be supported by the information system. It means that the contents of the information system cannot be based on the static Real World rules only. It has to be able to support primarily the intentions of its users. Intentions of the users of the information system are naturally changing and therefore they should not be hard wired in the information system with the standard, typified functionality. Instead of it the information system has to be permanently able to accommodate its behavior to the immediate needs of business processes.

Figure 4 illustrates the idea of the information system of a process-driven organization. It consists of the database and the predefined functionality which covers all standard business activities which are intended to be supported with the information system. System functions use the data from the system inputs and the database to transform them to the system outputs and transformed data in the database. System database contains the actual information about the state and the relevant history of the Real World. This part of the information system can be regarded as static as it represents the predefined, relatively stable and unchanging functionality. The dynamics of the information system required by the business processes (see the upper part of the figure) is ensured by the standard component 'Workflow Management System' (WMS) which allows combining the standard functionality of the system according to the needs of processes. As business processes are naturally dynamic, still changing, the WMS also uses the system database for storing the data about processes, their states and other attributes. The most important part of the WMS thus must be the tool for creating and changing the descriptions of processes which ensures the real dynamics of the whole information system. This way the business processes are supported by the control data and business rules which support their run as well as by the functional data

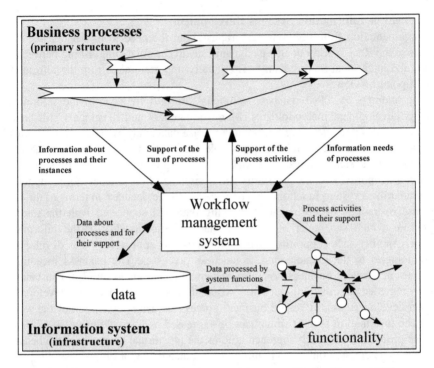

Fig. 4. Information system of a process-driven organization (Source: [14])

and business rules which support their activities. Business processes then deliver the information about themselves (changes in process definitions, state information and performance information) and about their information needs (requirements for functions).

4   Conclusions

The purpose of this paper is to emphasize the most important challenges in the field of business system modeling with use of the Philosophical Framework for Business System Modeling as a tool for the comprehensive understanding all important aspects of the Real World which should be reflected in the information system and conse-quently in information systems development methodologies. The used approach is build on the belief in the critical importance of the Principle of Modeling which makes the modeling of the Real World the most critical part of the information system development process. Using the Real World as an objective determinant allows us to compare its natural aspects to the state of the art in the field of information systems development and this way uncover the most important challenges usually in the form of current methical insufficiencies. Therefore, this paper is often critical to the men-tioned methods, languages and approaches.

The paper discusses particular detail consequences of the nature of the Real World from the point of view of different research areas: business process modeling, conceptual modeling and information system development.

Besides those detailed consequences the framework also shows that the common opinion that the main difference between the conceptual model and the process model is a difference between the static (conceptual model) and the dynamic (process model) view on a business system is a common mistake. Process model always expresses the intentional dynamics (behavior) while object-oriented models can also describe dynamics but with different meaning: not a behavior but a causality. So the real difference between the process- and the object (conceptual)-oriented models is the difference between the intentionality and modality. The difference between static and dynamic description of the Real World is the difference between the system and the particular views. This finding can be regarded as a common general challenge for the current state of the art in the field of information systems development.

Acknowledgments. The paper was processed with contribution of long term institutional support of research activities by Faculty of Informatics and Statistics, University of Economics, Prague.

References

1. Business Process Model and Notation (BPMN). OMG Document Number: formal/2011-01-03 (2011). Standard document URL: http://www.omg.org/spec/BPMN/2.0
2. Codd, E.F.: Recent investigations into relational data base systems. IBM Research Report RJ1385, 23 April 1974. Republished in Proc. 1974 Congress, Stockholm, Sweden, 1974. North-Holland, N.Y. (1974)
3. Codd, E.F.: A relational model of data for large shared data banks. Commun. ACM **13**(6), 377–387 (1970)
4. Emerson, E.A.: Temporal and modal logic In: Handbook of Theoretical Computer Science, vol. B, pp. 995–1072. MIT Press, Cambridge (1990)
5. Eriksson, H.E., Penker, M.: Business Modeling with UML: Business Patterns at Work. Wiley, USA (2000)
6. Guizzardi, G.: Ontological Foundations for Structural Conceptual Models. Telematica Instituut Fundamental Research Series No. 15 (2005). ISBN 90-75176-81-3, ISSN 1388-1795
7. Hammer, M., Champy, J.: Reengineering the Corporation: A Manifesto for Business Revolution. Nicholas Brealey Publishing, London (1993)
8. Heller, B., Herre, H.: Ontological categories in GOL. Axiomathes **14**, 71–90 (2004). Kluwer Academic
9. Chen, P.P.-S.: The entity-relationship model-toward a unified view of data. ACM Trans. Database Syst. **1**(1), 9–32 (1976)
10. Mayer, R.J., Menzel, C.P., Painter, M.K., deWitte, P.S., Blinn, T., Perakath, B.: IDEF3 Process Description Capture Method Report, Knowledge Based Systems, Inc. (1997)
11. Rosenblueth, A., Wiener, N., Bigelow, J.: Behaviour, purpose and teleology. Philos. Sci. **10**, 18–24 (1943)

12. Repa, V.: Business process modelling notation from the methodical perspective. In: Cezon, M., Wolfsthal, Y. (eds.) ServiceWave 2010. LNCS, vol. 6569, pp. 160–171. Springer, Heidelberg (2011). doi:10.1007/978-3-642-22760-8_18
13. Repa, V.: Modelling life cycles of generic object classes. In: Linger, H., Fisher, J., Barnden, A., Barry, C., Lang, M., Schneider, C. (eds.) Building Sustainable Information Systems, pp. 443–454. Springer, Boston (2013). doi:10.1007/978-1-4614-7540-8_34
14. Řepa, V.: Technical consequences of the nature of business processes. In: Gołuchowski, J., Pańkowska, M., Linger, H., Barry, C., Lang, M., Schneider, C. (eds.) Complexity in Information Systems Development. LNISO, vol. 22, pp. 185–199. Springer, Cham (2017). doi:10.1007/978-3-319-52593-8_12
15. Scheer, A.W: Architecture of Integrated Information Systems: Principles of Enterprise Modeling. Springer, Berlin (1992)
16. UML Superstructure Specification, v2.4.1, OMG Document Number: formal/2012-05-07, April 2012

Contextual Factors of Architectural Strategy for Complex Systems

Mirjana Maric$^{(\boxtimes)}$, Predrag Matkovic, Pere Tumbas,
and Veselin Pavlicevic

Faculty of Economics in Subotica, University of Novi Sad,
Segedinski put 9-11, 24000 Subotica, Serbia
{mirjana.maric, predrag.matkovic, pere.tumbas,
pavlicevic}@ef.uns.ac.rs

Abstract. Architecture is the "backbone" of every software product, regardless of the development process used. However, its role, significance, and development strategies differ from one software development process to another. Traditional architecture development, based on a well-defined architectural process that involves the three following architectural phases–architectural analysis, synthesis, and evaluation–is based on the Big Design Up Front strategy. In agile development, architecture is generated gradually with each iteration, as a result of continuous code refactoring, not some predefined structure. Therefore, agile software development relies on an opposite extreme architectural strategy, emergent architecture.

The research topic of this paper is focused on the development of architecture for modern complex systems, which cannot be based on either of the two aforementioned extreme architectural strategies. Development of an architectural strategy for a complex system is significantly influenced by contextual factors.

This paper presents the results of a qualitative empirical research, carried out on a sample of 20 expert practitioners. The results represent contextual factors that the practitioners–surveyed respondents–consider when deciding to which extent will the emergent strategy be extended with explicit architectural practices typical to the traditional architecture development. Obtained results suggest that agile practitioners scale up agile processes, in terms of architecture development, with regard to the contextual factors of the system being developed.

Keywords: Agile software development · Software architecture · Contextual factors

1 Introduction

Agile development processes are nowadays used by countless software development companies worldwide. The main motivation for the use of agile processes is to reduce costs and increase adaptability to changes resulting from dynamic market conditions [1]. Emergence of agile development processes has had a significant effect on the software industry, but it also opened numerous issues that are currently occupying academic researchers. One of the current questions pertains to the role of software architecture and its significance in agile processes [2, 3].

© Springer International Publishing AG 2017
S. Wrycza and J. Maślankowski (Eds.): SIGSAND/PLAIS 2017, LNBIP 300, pp. 111–123, 2017.
DOI: 10.1007/978-3-319-66996-0_8

Software architecture implies decomposing the system into smaller parts, modules, which are developed independently and can be reused or replaced with other modules. Modularization increases flexibility by creating loose links between highly interconnected parts of a system, localizing the effect of changes made to a particular component; it improves the comprehensibility of the system, easing its maintenance and further development [4].

There are two opposite strategies for developing software architecture, with completely different attitudes to the role and the significance of architecture in the software development process: emergent architecture and Big Design Up Front (BDUF).

BDUF strategy involves a well-defined architectural process, comprised of a vast number of explicit architectural activities and decisions that precede the implementation of system functionalities. In other words, the complete architectural design is done at the beginning of a project. This architectural strategy is typical to the traditional software development.

In contrast, emergent architecture is typical to agile development, guided by the principles of "value driven development" and focused on early delivery of value to the user. Agile processes emphasize the value of early development of functionalities, while seeing the architecture as an outcome of the development process. Agile practitioners often consider architecture development economically unjustified, believing that it does not provide ROI, but rather increases total project costs. In agile processes, concepts of metaphor and refactoring are considered adequate replacements for the traditional architecture development process. More exactly, agile processes do not have typical architecture development activities, such as analysis, synthesis, and evaluation [5]. Architecture is rather developed gradually with every iteration, as a result of continuous changes to the source code, not as a result of a predefined structure [5–7].

However, development of architecture of modern complex systems must be based on a strategy that balances the previously described extremes. It would be highly risky to rely entirely on emergent architecture in complex system development, since it is possible to reach a point where the present architecture cannot be amended through refactoring. Such situation would require a complete redesign of the architecture, which implies a significant increase in costs, along with client dissatisfaction. Although agile processes help developers achieve efficiency, quality, and flexibility in change management, complex system development requires application of explicit architectural practices [8–11]. This helps avoid high amount of refactoring, and reduces the risk of architectural erosion [11–13].

The extent to which explicit agile architectural practices are included in an agile development process depends on the contextual factors of the system. In accordance with the described research subject, the research problem addressed in this paper is the influence of contextual factors on architectural strategies for complex system development.

The research question was defined in line with the defined research problem:

RQ: Which empirically identified contextual factors influence the application of explicit architectural activities and the extent of up-front architectural planning in agile software development processes?

The answer to this research question will be given in an overview of primary qualitative results of the conducted empirical research.

2   Research Methodology

This empirical research was based on qualitative methods. Collection of empirical data was conducted by means of a semi-structured interview. The initial set of questions was based on an analysis of prominent literature related to the research topic. The research instrument was subject to evaluation by a group of experts. Subsequent to expert evaluation, content validity index was computed for each of the questions, as well as for the entire questionnaire, in accordance with the recommendations by Polit and Beck [14]. The content validity index of the first version of the questionnaire was 0.76, which suggested that it needed to be modified in accordance with expert's suggestions. Modifications involved elimination of questions with values of the content validity index below 0.8, amendment of how certain questions were formulated, as well as merging several questions into single questions.

The interviews were conducted face-to-face, recorded, transcribed and submitted to interviewees for verification. Research results were obtained through thematic content analysis. Coding of interview data and the thematic content analysis were carried out in the NVivo software suite, following recommendations by Miles and Huberman [15]; all interview transcripts were studied several times prior to the development of the initial list of codes, which was based on the relevant literature; new codes were introduced inductively throughout the analysis. Codes were grouped into categories based on similar characteristics and subsequently organized into clusters.

The nature of the research necessitated purposive sampling (n $\geq$ 20). Therefore, the sample consisted only of experts with significant experience (minimum 5 years) in complex system development with agile processes and software architecture development, selected from prominent Serbian IT companies. Since the study was limited to one country, the authors purposely included respondents outsourced by companies from various countries, as well as ones employed by global software development organizations. Bootstrapping with 1000 replications was carried with the aim to increase the stability of the findings.

3   Contextual Factors of Software Development

Most of the software development processes commonly recognized as "agile", such as Scrum and XP, have now been around for more than two decades. The term "Agile" denotes iterative and incremental approach to software development, which relies on the principles proclaimed in the Agile Manifesto [16]. Over the years, there has been much interest in the architectural implications of applying agile processes in large-scale projects. Kruchten [17] introduced the term "Agile Sweet Spot" to designate the conditions from which the agile methods stemmed, are where they are most likely to succeed. The author states that out-of-the-box agile methods are most suitable for small, collocated groups of 10–15 people (which allows for good communication), sharing a common culture and working in close relations with the customer on a new ("green field") software project with s short lifecycle. As argued by Abrahamsson et al. [2], a project in such settings does not require many architectural activities. However, as the project context moves from the "agile sweet spot", the process needs to embrace certain architectural practices, suited to the project context.

Several authors have made efforts to classify the contextual factors of software development projects. In a paper concerning the applicability of agile development practices in projects with different characteristics and backgrounds, Kruchten [18] presented empirically identified contextual factors of software development, divided the into two levels: (1) organizational level (environmental) factors, and (2) project-level factors. The organizational level factors include the business domain, number of instances of the software system to be deployed, maturity of the organization, and the level of innovation. Project-level factors (influenced by organizational level factors) include: system size, stable architecture (whether there already is a set of commonly used patterns), business model, team distribution, rate of change, age of system, criticality, and governance. The latter constitute what the author referred to as the "octopus model" of key contextual factors of software development, depicted in Fig. 1. Such classification of factors enables teams to carry out an analysis of the project context prior to its initiation and determine which practices or methods are unsuitable, and which are essential.

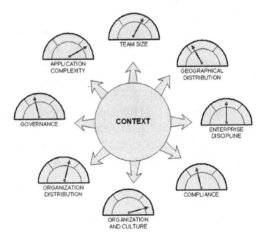

Fig. 1. Contextual factors for agile software development (adapted from [18])

Similar to Kruchten et al. [18], Ambler [19] composed a set of scaling factors that influence the efforts of applying agile and lean processes to large projects: (1) team size – from under 10 developers to hundreds of developers; (2) geographical distribution – from co-located to globally distributed; (3) compliance – from low risk to critical/audited; (4) organization and culture – from open to entrenched; (5) organization distribution – from in-house to third party; (6) governance – from informal to formal; (7) application complexity – from simple, single platform to complex, multi-platform; (8) enterprise discipline – from project focus to enterprise focus. These factors are intended as a reference in the analysis of the context prior to tailoring the development process to suit an individual project.

Boehm et al. [20] have devised a set of guidelines for finding the right balance between agility and architecture in the type of projects that require both. They focus on three contextual factors for determining how much agility and architecting are enough in a software development project [21]: project size, criticality, and volatility. Babar et al. [22] assumed that the three factors proposed by Boehm et al., coupled with other possibly influential contextual factors, also determine whether a satisfactory architecture can emerge from continuous refactoring in an agile project. Their empirical research resulted in a framework with 20 contextual factors, subsequently grouped into four distinct categories: project, team, practices, and organization.

As the project domain shifts from the "agile sweet spot", low ceremonialism and high interactivity, the key characteristics of agile processes, seem to increase the risk of project cost proliferation, architectural erosion, or even project failure. In the following text, we discuss the results of the empirical research, where we aim to identify contextual factors used by practitioners to determine the extent of up-front architectural analysis and design necessary for a particular project.

4   Research Results

Qualitative analysis of the empirical data through inductive text coding resulted in key categories, which served as a structure for the presentation of research results. The main category, titled "Agile Architecture Design" concerns the very process of practical agile architecture development. This category comprises three subcategories, namely: "Factors od Architectural Strategy", "Explicit Architectural Activities", and "Roles and Responsibilities in Architectural Decisions". All of the subcategories influence the development of agile architecture of complex systems in a specific sense; however, in line with the research questions, this paper will be focused on the category "Factors of Architectural Strategy".

The category "Factors of Architectural Strategy" contains numerous concepts, further divided into four subcategories:

1. System complexity: complexity of the problem being solved, complexity of the problem domain, complexity of the architectural solution, complexity of requirements
2. Requirements: quality and volatility
3. Team characteristics: experience, communication, knowledge, team size, etc.
4. Stakeholders: agility, quality of communication, knowledge, etc.

The subcategories listed above represent empirically identified contextual factors that agile practitioners take into consideration when developing architectural strategies for complex system development. Architectural strategies for complex systems always fall between the two extremes: emergent architecture, and BDUF. Which extreme will it be closer to depends on the contextual factors, which are presented and clarified in the following text, along with the interviewees' responses (RSP 1-20).

4.1 System Complexity

System complexity is determined by three main attributes: scale (number of things within a system), diversity (variety of things in a system), and connectivity (number of relationships between the things) [23]. Agile teams have explained complexity through the following terms: complexity of the problem being solved, complexity of the problem domain, complexity of requirements and the architectural solution.

Respondents believe that the complexity of the problem being solved and the complexity of the problem domain largely influence the length of the planning phase, especially in cases when the problem/domain is vast and cannot entirely be comprehended by a single person, but requires brainstorming with multiple team members, who are experts in their respective areas (RSP16).

Complexity of the architectural solution has a significant impact on the architecture development strategy as well. It determines the scope of architectural decisions that need to be made prior to implementation of system functionalities. This factor is most influential in mission safety critical products, where even the slightest problems with system performance can cause disastrous consequences.

In practice, greater complexity of requirements also increases the scope of up-front architectural decisions, which moves the architectural strategy towards the BDUF extreme. Complex system requirements can include cross-platform compatibility, integration with a legacy system, alignment with the legislation, or fault tolerance. These requirements represent system constraints and imply a greater extent of architectural analysis and planning at the beginning of a project, each release, or even a sprint (RSP14).

Requirement for integration and interaction with legacy systems demands spending considerable time on the analysis of legacy systems and the technology used for their development. The challenge of understanding legacy systems becomes even greater if the documentation of the legacy system is sparse, or even non-existent. Similarly, finding the best way to interact with the legacy system is no less of a challenge (RSP1, RSP19). Respondents stated that the biggest problem with this type of requirements is to resolve all issues originating from legacy data, i.e., to determine which data is relevant to the development of the new system's data model, where the legacy data will be stored, and how they would be managed until the software is in production. Transition from a legacy system is a secondary project that should be planned within the development project (RSP15).

According to the respondents' experience, the requirement of cross-platform compatibility has a vast influence on the architectural solution. Therefore, it necessitates in-depth analysis at the beginning of the project, aimed at selecting adequate technology. The amount of time spent on determining the right technology greatly depends on the software architect's knowledge and experience. Choice of technology is an important moment in architecture development, since not all technologies are equally suitable for solving particular problems and developing certain types of software products (RSP3, RSP5, RSP20).

The complexity of requirements is also determined by the number of users, number of transactions, as well as the amount of data that the system is expected to process. These requirements are directly related to scalability, and therefore must be thoroughly considered prior to implementation (RSP13, RSP17, RSP2).

It is interesting that the respondents did not directly relate project size with complexity, which contravenes the findings in the literature [2, 24, 25]. The respondents believe that the fact that a project requires a great amount of time, and/or participation of many people does not imply a greater extent of architectural planning at the start. A large-scale project may just require a lot of "manual labor" on coding and the development of the solution (RSP12). The respondents associated the scale of a project with project management challenges, which can purportedly be mitigated through service or microservice architectural solutions (RSP16).

The described empirical results point to the conclusion that the respondents directly link the factors of complexity with the extent of up-front architectural decisions necessary in architecture development. In other words, the greater complexity of a system being developed will cause the emergent architecture to be based on a greater extent of up-front architectural decisions. It should also be noted that architecture development essentially remains incremental, eliminating the possibility of falling into the trap of BDUF. The respondents stated that the time necessary for up-front architectural decisions can significantly be reduced with the use of a reference architecture (predefined architectural solutions). Such respondents' opinions are in line with the findings in the literature [26, 27], which suggest that the use of a reference architecture improves the agility of the development process.

System requirements, especially their volatility and insufficient quality, significantly influence the choice of an architectural strategy and its positioning in relation to one of the two extremes.

4.2 Requirements

Most of the respondents stated that projects are often realized in unstable conditions, since the clients are not entirely sure what they need at the beginning of a project. They mostly come with an idea that they had not thoroughly elaborated, which can result in wearisome and lengthy identification of architectural requirements that can delay the implementation phase. In addition to that, clients are often unable to clearly formulate their requirements, which is yet another cause of delay (RSP8, RSP10, RSP6).

Quality of requirements is in direct proportion to the amount of time a software architect needs to spend on up-front architectural analysis at the beginning of a project. Identification of architecturally significant requirements at the beginning of the project is beneficial not only to the development team, but to the client as well. Architecturally significant requirements serve as the basis for determining the scope of the future system, and are crucial to the design of the architecture for the principal part of the software (RSP8, RSP4). This is where non-functional requirements, of which the clients are mostly unaware, are of utmost importance. For this reason, the software architect should help their clients identify non-functional requirements and discuss the possibilities of their implementation, since some non-functional requirements may be mutually exclusive (RSP3, RSP4).

Initial definition of architecture involves identification of elements that are costly to develop, and therefore should not undergo changes throughout the project. Subsequent identification or modification of these crucial architectural raises the question whether to develop an entirely new architectural design, or to attempt refactoring the present unsatisfactory solution. In both cases, costs increase enormously, as well as the duration of the project (RSP8).

In order to counteract these challenges, practitioners often use the concept of spikes. Spikes are used in two different cases: when the clients have a certain requirement, but are unsure what they want in terms of functionalities, and when the software architect and the development team are unsure how to implement certain requirements. In these cases, practitioners most frequently opt for developing a time-boxed architectural prototype.

Volatility of requirements is the second most common problem software architects and agile teams face. Volatility of requirements is associated with the implementation stage, when the client states changes to their expectations from the software. This can severely jeopardize the viability of the developed architecture (RSP13, RSP12).

This risk can be mitigated through additional efforts at the beginning of the project, aimed at reaching a consensus with the stakeholders on the business and architectural vision. This does not involve a detailed elaboration of all the requirements, which would not even be possible in the modern-day business environment, but rather focusing on architecturally significant requirements for the main part of the system. Nowadays, software is crucial to the success of a business. In line with this, software development is a continuous process, as is any business process within an enterprise. Changes in business instigate new requirements and continuous adaptation of the software solution, which implies that architecture development must be a continuous, iterative and incremental process. The architecture initially developed for the main part of the system evolves throughout the project, while the detailed elaboration of a set of requirements takes place during the planning stage of an iteration in which they will be implemented. This lead to a conclusion that numerous architectural decisions must be postponed until the requirements are well understood, which can be called Just-In-Time architectural planning (RSP6, RSP9).

Changes to the initially set architecture is justifiable only in case of a great advance in the client company's operations and a much more extensive use of the system than the intended, when the performance of the present architectural solution acts as a bottleneck (RSP8).

These results suggest that the respondents believe that there is a direct link between the quality and volatility of requirements and the length of the architectural planning stage. Quality of requirements influences the amount of time a software architect needs to spend on up-front architectural analysis, in order to determine the scope of the system and architecturally significant requirements, which serve as the basis for the architecture of the main part of the software.

4.3   Stakeholders

According to the respondents, stakeholders' qualities represent an important factor of architectural strategy definition. Their openness to continuous collaboration through active participation in the project and delivery of feedback to the project team influences the choice of architectural strategy. Respondents' views are in line with the findings in the literature; specifically, Friedrichsen (2014) pointed that the extent of up-front analysis and design depends on software architect's experience, skills, knowledge, as well as good communication with the stakeholders.

In addition to stakeholders' openness to active collaboration throughout the project, the quality of their involvement and communication with them is also an important factor. Stakeholders should understand the problem, and also be able to formulate and express their requests and needs (RSP4, RSP13).

A great portion of problems in a project, including those associated with software architecture, originates from stakeholders' lack of competence to provide adequate requirements. Therefore, for the success of the software architecture and the whole project, it is of utmost importance to identify key stakeholders, i.e. the stakeholders with the greatest influence on the software being developed, at the start (RSP7, RSP8, RSP13).

The choice of an architectural solution should be a result of a close collaboration between the software architect and the stakeholders. The software architect's role is to present stakeholders with alternative architectural solutions and descriptions of expected results each of the solutions would provide (RSP3), along with the costs associated with each solution (RSP8). The stakeholder's role is to make the final decision on the course to be taken in the architecture development (RSP3).

It is challenging, even impossible, to follow an architectural strategy close to emergent architecture when working for a non-agile stakeholder (client/user). Stakeholders often refuse to accept the principles of agile development, primarily because they want to "insure" themselves by defining the scope, timeline and quality in the contract. If all this is preset, it is not possible to develop software with full adherence to agile principles (RSP9). This situation is particularly evident in complex system development projects, where clients tend to have a negative attitude towards fully agile development. This is principally because of the elimination of the planning stage and emergent architecture, distinctive of agile processes (RSP8).

Based on all above, it can be concluded that stakeholders' qualities are a factor that influences the extent of architectural planning in complex system development. The results suggest that stakeholders do not support architectural strategies based entirely on emergent architecture.

4.4   Team Characteristics

Results within this category suggest that characteristics of a development team, such as team size, understanding of the problem being solved, knowledge in the problem domain, familiarity with technologies and current trends in architectural options, experience, and quality of communication and collaboration influence the development of complex systems' architecture with agile processes

If the team was involved on solving a similar problem before, then they possess relevant experience and knowledge, which reduces the time and effort required for up-front architectural analysis and design. This implies that software architects do not need to develop prototypes or organize brainstorming sessions, since they can decide on the architectural solution based on their experience (RSP11).

Team member's familiarity with the problem domain, as well with the technology to be used in the development of a solution also reduce the time and effort required for up-front architectural analysis and design (RSP9, RSP3). If the team is composed of individuals with experience with different alternative technologies and familiarity with their capabilities and limitations, up-front architectural planning and the choice of technology will be accelerated. Nowadays, there are countless free components available, which enables faster development and lower costs (RSP2, RSP5). However, situations where clients impose their technology stack are frequent, especially in outsourcing projects. In such situations, the team requires additional time to explore the technology, as well as to consult with the individuals of enterprises with experience in using these technologies. All of this delays the implementation of functionalities (RSP14, RSP3, RSP18).

In addition to that, time invested in up-front architectural activities will be significantly shorter if the team is composed of individuals with proper knowledge and experience in the domain of software architecture. Experienced software engineers can make a significant proportion of architectural decisions unaided by the software architect, which increases the agility of the team. The younger and less experienced team members are, the more time the software architect needs to spend on up-front architectural planning and design, as to trace out the developers' work as much as possible and mitigate the risk of taking a wrong turn in the development (RSP16, RSP5).

In order to gradually reduce the time and effort the software architect needs to invest at the beginning of the project and during the implementation of the solution, it is necessary to commit to continuous advancement of all team members' knowledge. This practically means that all team members should attend software architects' meetings and actively participate in architectural planning and development of architectural strategies. In addition to that, they should undergo courses such as "professional scrum developer", etc. Education and communication are, therefore, means of enabling developers to unassistedly make architectural decisions during the implementation stage (RSP14).

Such respondents' views correspond to the findings in the literature. Boehm and Turner [20] have concluded that agile development requires a critical mass of experts. The authors defined experts as the members of the team "Able to revise a method (break its rules) to fit an unprecedented new situation." An expert should not blindly follow the rules and instructions, but to have enough knowledge, skills and experience to be able to make decisions based on intuition. Blair, Watt i Cull [28] highlighted the necessity of close cooperation between the software architect and the team, with continuous sharing of ideas and knowledge throughout the entire project.

Most respondents have recognized the number of teams/individuals involved in a project as another factor influencing the extent of up-front architectural planning, particularly owing to the necessity of analyzing dependencies among teams (RSP4). If the components being developed separately by different teams are interconnected, this implies a completely different approach to architectural planning and coordination, as well as to implementation and testing.

Appropriate communication and collaboration is the only way to reduce the need for greater up-front planning in case of a large number of teams.

Respondents' opinions on this issue correspond with the conclusions of Coplien and Bjornvig [29], who stated that architectural efforts can radically be reduced through good communication between team members, also providing an example of reduction from six months to two weeks.

However, agile collaboration, good communication, and experience are things that agile teams cannot achieve overnight, but rather require that team members spend time working together on various projects. The longer the team members spend together, less time is required for up–front architectural planning, since there are fewer problems in the communication between the developers and software architects. In such cases, it is sufficient to agree on the patterns to be used, discuss the specificities of particular aspects of development, and identify critical relations at the very beginning of a project. The goal is to elevate developers' skills and knowledge in the domain of software architecture, making them able to "tailor" parts of software by themselves, without detailed instructions by an architect (RSP19).

This is in line with empirical findings of Hoda [30], who determined that it requires a certain time for members of a team to develop experience, self-organization, self-evaluation, and self-improvement.

In addition to that, an agile team, in the real sense of the word, implies members sitting in the same room and communicating face-to-face on a daily basis. This further means that agile processes were intended for in-house development projects. However, the respondents stated that most of their projects involve outsourced work, which is the principal reason that these projects can not entirely rely on agile principles and values, but rather need to be scaled through inclusion of explicit architectural activities (RSP14).

5 Conclusion

Research results suggest that agile teams involved in the development of complex systems do not use either of the two extreme strategies for developing software architecture–BDUF and the emergent architecture–but rather employ strategies that fall between the two extremes. Contextual factors of the system being developed determine to which extreme the selected architectural strategy will be closer. This is a matter of establishing a balance between the tactical level of software development, one that provides rapidly visible value through the development of functionalities, and the strategic level, which produces long-term value through the design of software architecture.

Research results point to a conclusion that empirically identified contextual factors (system complexity, system requirements, stakeholders' qualities, and team characteristics) influence the time required for architectural planning in complex system development. Even though this phase is almost entirely eliminated in typical agile projects, results of this research indicate that this is not the case in the development of systems, which corresponds with the views found in previous studies. The reason behind this lies in the fact that agile processes were not initially intended for complex system development, and therefore emergent architecture is an inadequate strategy when this type of software is concerned. In order to be applied in complex system development, agile processes need to be scaled up through inclusion of explicit architectural activities, typical to the traditional architecture development. In other words, the authors of this study recommend extending the outreach of architectural planning beyond the current sprint, as to prevent loss of flexibility and degeneration of the design. The scope of explicit architectural activities carried out at the beginning varies from project to project, since the contextual factors that influence them are unique for each system. Therefore, the authors recommend a detailed analysis of contextual factors of the future system be carried out prior to the definition of an architectural strategy.

It can be concluded that the development of a complex systems' architecture based entirely on emergent architecture may not necessarily result in "agile architecture". The term "agile architecture" designates an architecture designed in a way that it can easily be changed, i.e., that it can react to the changes in the environment. Therefore, the authors recommend retaining an iterative process for developing complex systems' architecture, but one comprising an adequate number of explicit architectural practices, executed Just-In-Time, and in line with the contextual factors of the future system.

References

1. Ambler, S.W., Lines, M.: Disciplined Agile Delivery, 1st edn. IBM Press, Boston (2013)
2. Abrahamsson, P., Babar, M.A., Kruchten, P.: Agility and architecture: can they coexist? IEEE Softw. **27**(2), 16–22 (2010)
3. Pérez, J., Díaz, J., Garbajosa, J., Yagüe, A.: bridging user stories and software architecture: a tailored scrum for agile architecting. In: Babar, I., Brown, M.l., Mistrik, A.W. (eds.) Agile Software Architecture: Aligning Agile Processes and Software Architectures, 1st edn., pp. 215–241. Elsevier, New York (2014)
4. Parnas, D.L.: On a 'Buzzword': hierarchical structure. In: Proceedings of the IFIP Congress 1974, pp. 336–339 (1974)
5. Babar, I., Brown, M.l., Mistrik, A.W.: Agile Software Architecture Aligning Agile Processes and Software Architectures, 1st edn. Elsevier, Waltham (2014)
6. Beck, C., Andres, K.: Extreme Programming Explained: Embrace Change, 2nd edn. Addison Wesley, Boston (2004)
7. Thapparambil, P.: Agile architecture: pattern or oxymoron? Agil. Times **6**(1), 43–48 (2005)
8. Babar, M.A., Abrahamsson, P.: Architecture-centric methods and agile approaches. In: Proceedings of the 9th International Conference on Agile Processes and eXtreme Programming in Software Engineering, pp. 238–243 (2008)
9. Parsons, R.: Architecture and agile methodologies—how to get along. In: WICSA (2008)
10. Nord, R.L., Tomayko, J.E.: Software architecture-centric methods and agile development. Softw. IEEE **23**(2), 47–53 (2006)
11. Ihme, T., Abrahamsson, P.: The use of architectural patterns in the agile software development on mobile applications. In: ICAM 2005 Internetional Conference on Agility, vol. 8, pp. 1–16 (2005)
12. Stal, M.: Refactoring software architectures. In: Agile Software Architecture: Aligning Agile Processes and Software Architectures, pp. 130–152. Elsevier (2014)
13. Kruchten, P.: Situated agility: context does matter, a lot. In: 9th International Conference on Agile Processes and eXtreme Programming in Software Engineering (2008)
14. Polit, C.T., Beck, D.F.: The content validity index: are you sure you know what's being reported? critique and recommendations. Res. Nurs. Heal. **29**, 489–497 (2006)

15. Miles, M.B., Huberman, A.M.: Qualitative Data Analysis: An Expanded Sourcebook, 2nd edn. Sage, Thousand Oaks (1994)
16. Beck, K., et al.: The Manifesto for Agile Software Development (2001)
17. Kruchten, P.: Scaling down large projects to meet the agile 'sweet spot'. Ration. Edge, 1–14, August 2004
18. Kruchten, P.: Contextualizing Agile Software Development, pp. 1–12 (2010)
19. Ambler, S.W.: Agility@Scale: Strategies for Scaling Agile Software Development. Agile and domain complexity (2010)
20. Boehm, R., Turner, B.: Balancing Agility and Discipline: A Guide for the Perplexed, 1st edn. Addison-Wesley, Boston (2003)
21. Boehm, B., Lane, J., Koolmanojwong, S., Turner, R.: Architected agile solutions for software-reliant systems. In: Agile Software Development: Current Research and Future Directions, pp. 165–184. Springer, Heidelberg (2010)
22. Chen, M.A., Babar, L.: Towards an evidence-based understanding of emergence of architecture through continuous refactoring in agile software development. In: 2014 IEEE/IFIP Software Architecture (WICSA), pp. 195–204 (2014)
23. Kruchten, P.: Complexity made simple. In: Proceedings of the Canadian Engineering Education Association (2012)
24. Boehm, R., Turner, B.: Using risk to balance agile and plandriven methods. IEEE Comput. 35(6), 57–66 (2003)
25. Cockburn, A.: Agile Software Development: The Cooperative Game, 2nd edn. Addison Wesley, Boston (2007)
26. Mirakhorli, J., Cleland-Huang, M.: Traversing the twin peaks. IEEE Softw. 30(2), 30–36 (2013)
27. Kruchten, J., Obbink, P., Stafford, H.: The past, present, and future for software architecture. IEEE Softw. 23(2), 22–30 (2006)
28. Blair, S., Watt, R., Cull, T.: Responsibility-driven architecture. IEEE Softw. 27(2), 26–32 (2010)
29. Coplien, G., Bjornvig, J.O.: Lean Architecture for Agile Software Development, 1st edn. Wiley, Cornwall (2010)
30. Hoda, R.: Self-Organizing Agile Teams: A Grounded Theory. Victoria University of Wellington, Wellington (2010)

Investigating Enterprise System Adoption Failure Factors from an Employee Age Perspective

Ewa Soja[1](✉) and Piotr Soja[2]

[1] Department of Demography, Cracow University of Economics,
Kraków, Poland
Ewa.Soja@uek.krakow.pl
[2] Department of Computer Science, Cracow University of Economics,
Kraków, Poland
Piotr.Soja@uek.krakow.pl

Abstract. The current study seeks to investigate the role of employees' age in the perception of reasons of failure of enterprise system (ES) implementations. On the basis of empirical data gathered from Polish ES practitioners and following grounded theory approach, the taxonomy of reported failure factors has been developed. Next, an analysis into the differences of failure factors perceptions depending on respondent age has been conducted. The age-based analysis illustrated that perceptions of ES failure factors might vary with employee age. The main findings suggest that older employees tend to perceive ES failures from the perspective of a company and its organization, while younger workers recognize causes of ES failures mainly in the organization of ES implementation project.

Keywords: Enterprise systems · Adoption · Failure factors · Labor force age · Ageing · Poland

1 Introduction

Enterprise systems (ES), having their roots in MRP, MRP II, and ERP systems, are now very complex systems that support the management and integration of the whole company and offer inter-organizational integration with company's clients and suppliers [28]. Although ES can bring many potential benefits to adopting organizations, its implementation is a risky and challenging process. In consequence, many organizations fail to realize the expected benefits and their ES adoption projects considerably exceed estimated budgets and schedules (e.g. [1, 31]). Therefore, undoubtedly, investigating reasons of ES adoption failures appears an important and promising research topic.

ES adoptions engage many stakeholders both from within the adopting company and external organizations [18]. In consequence of a full-fledged ES implementation within an organization, practically all employees of the organization are affected by the ES project. Demographic characteristics of the employees might be highly diversified due to workforce ageing – a phenomenon experienced by developed societies [2, 15].

© Springer International Publishing AG 2017
S. Wrycza and J. Maślankowski (Eds.): SIGSAND/PLAIS 2017, LNBIP 300, pp. 124–135, 2017.
DOI: 10.1007/978-3-319-66996-0_9

Older employees differ from their younger counterparts in work capacity and attitudes toward new technologies. In particular, as age increases attitudes towards computers tend to be more negative and technology anxiety increases [12, 29]. Older workers are perceived by employers as having a higher knowledge base, but also generating higher costs and revealing lower productivity levels (e.g. [5, 27]).

In view of widely reported issue of work force ageing we would like to investigate whether employees' age matters in the perception of ES adoption failure factors. In particular, we would like to answer the following research questions:

1. What are the factors that might cause ES adoption failure?
2. Did employees' age play any role in the perception of ES adoption failure factors?

The paper is organized as follows. In the next section we present research background concerning implications of ageing and factors that might contribute to failure of ES adoption projects. Next, we describe our research method, which is followed by the presentation of results. We then discuss our findings, explain implications, and close the study with concluding remarks.

2   Research Background

As already mentioned, the results of many ES implementation projects are not satisfactory and there are many failures among ES adoptions. Therefore, researchers conducted investigations in order to discover major reasons of ES failures. Prior research works involved various approaches to investigations and data gathering. There are studies which are conceptual propositions, research works drawing from secondary data, and studies based on primary data mainly gathered during case studies and interviews with experts. The following paragraphs shortly describe selected important studies analyzing ES failure factors.

Umble and Umble [25] posit that there are ten main categories of ES adoption failure causes. The categories include various issues associated with implementation project organization and run, condition of the adopting company, and involved technology. The proposed issues are:

- poor leadership from top management,
- automating existing redundant or non-value-added processes in the new system,
- unrealistic expectations,
- poor project management,
- inadequate education and training,
- trying to maintain the status quo,
- a bad match between an organization and the new system,
- inaccurate data,
- viewing ES implementation as an IT project,
- significant technical difficulties.

Gargeya and Brady [9] used the secondary data of SAP implementations in 44 companies and concluded that there are two primary factors that contribute to ES failure: inadequate internal readiness and training, and inappropriate planning and budgeting. Apparently, these factors refer to project organization and people involved in and affected by the project.

Wong, Scarbrough, Chau, and Davison [30] performed a multiple case study and interviewed different stakeholders including top management, project managers, project team members and consultants in order to investigate critical failure factors is ES implementation. Their findings imply that the most important critical failure factors refer to poor consultant effectiveness, project management effectiveness, and poor quality of business process re-engineering.

Chen, Law and Yang [4], based on a case study conducted in a US-based multinational organization, emphasize the role of poor project management as an important reason of ES adoption failure. In particular, the authors highlight the paramount role of risk management, communication, project scope management, and human resources during ES implementation project organization and run.

Hawari and Heeks [11] performed a case study in a Jordanian organization and concluded that ES adoption fail due to too large a gap between ES design and client organization reality; a gap that remains unclosed during implementation, and which exists on several dimensions. The most important dimensions include stakeholders' acceptance of objectives and values, and availability of experienced and skilled project staff. Other important dimensions refer to information processing, IT infrastructure, management systems and structures.

Amid, Moalagh and Ravasan [1], in order to examine ES critical failure factors, conducted interviews with Iranian ES project team members that failed in their projects. The authors concluded that ES failure factors might be grouped into seven categories related to vendor and consultants, human resources, managerial issues, project management, processes, organizational issues, and technical considerations.

Lech [13] focused his investigation on the implementation service provider side and conducted interviews with consultants and experts. The results suggest that the main failure factors refer to various issues associated with project management and project participants. The suggested project management-related failure factors are scope, budget, and schedule management, use of appropriate methodology, and good communication. The participant-related failure factors refer to competence and empowerment of project team members and expertise and involvement of provider's consultants.

Summing up the review of prior works dealing with ES project failure factors we can conclude that causes of ES failure might refer to various issues associated with project management, company reorganization, and people involved in the project. In consequence of ES adoption practically all company's employees are affected by the new system, therefore, the role of people-related considerations appears especially crucial for the project successfulness. This is particularly important in the light of ageing workforce, which is experienced by today's organizations and is predicted in the future (e.g. [2, 14]).

According to UN projections ([26], medium scenario) for Europe, the share of potential working age population will decrease by about 11% points between 2015 and 2060. At the same time, the labor force will gradually age. It is foreseen that the work force participation rate of older people (55–64) will increase by about 15% points between 2013 and 2060 [8].

Prior research suggests that work capacity changes along with employee age. For example, on the one hand, with age the ability to reason and comprehend the whole grow, as well as loyalty and faithfulness to employer rise. On the other hand, physical abilities decrease and reluctance to change and technology anxiety increase (e.g. [12, 27, 29]). Therefore, in order to be successful in the future, organizations in their strategies should take incorporate actions that would fully exploit the evolving capabilities of their employees.

The broad range of considerations experienced by ES adoption projects and many stakeholders involved in such projects suggests a vital need to examine ES projects from the perspective of multiple stakeholders. In particular, in light of an ageing society and ageing workforce it appears necessary to investigate ES adoption considerations from an employee age perceptive. Prior research in this field is scarce, with some studies employing an age-based perspective in the investigation of impediments to ES adoption [19, 22], areas for improvement of ES projects [20], and barriers encountered during ES adoption projects [21]. Overall, to the best of our knowledge, there is a gap in ES research related to the lack of an in-depth examination of employees' age and demographic background. Such an investigation appears worth studying since many ES adoption considerations are people-related and also because older employees are becoming more and more significant employee group in companies due to labor force ageing.

3   Research Method

In order to answer the involved research questions we employed a qualitative research approach based on grounded theory [3, 6, 10]. We turned to practitioners to learn what their opinions are as regards factors that might cause enterprise system adoption failure. In consequence, the current study is based on data gathered from practitioners dealing with ES implementation or involved in an ES package operation in various companies in Poland. In total, 29 practitioners expressed their opinions regarding failure factors in enterprise system adoption projects. The respondents generally participated in several ES projects and expressed their opinions based of their experience. The inquired respondents were diverse as regards their organizational position and roles played during the implementation process and represented both main parties involved in ES adoptions: adopter and provider.

During the process of data gathering, the respondents were asked to answer open-ended questions concerning the main causes of enterprise system adoption failures. In consequence of the data gathering process, respondent opinions expressed in natural language have been collected. Following the grounded theory approach, we then performed the process of open coding [6], where the respondent statements were compared and analyzed in the search of similarities and differences. The statements were given conceptual labels and categories were created. Next, the process of axial coding was performed, during which the relationships between categories which emerged during the process of open coding were tested against data and verified. As a result, the categorization of the reported failure factors was worked out and agreed upon by the authors.

The next step of the data analysis process was the age-related analysis, during which the discovered elements were analyzed taking into consideration respondent age group. In particular, drawing from prior demographic studies (e.g. [5, 23]) we defined our respondents into three age groups: the younger (less than 35 years old), the middle-aged (between 35 and 50), and the older (50+). During the age-related analysis, we analyzed the distribution of investigated categories across the defined age ranges.

4   Data Analysis and Results

4.1   Causes of Failure in Respondent Opinions

On the basis of empirical data analysis, fourteen main categories of failure factures have been extracted. They are described in the following paragraphs, ordered in decreasing level of importance in respondent opinions. It should be stressed here that the categorization of issues has been performed taking into consideration all respondents regardless of their age. The age-based examination was performed as the second next of the analysis step and is discussed in the next section.

- System – issues referring to the system being implemented, mainly associated with inadequate choice of the system and lack of system-organization fit, difficulties with data migration, poor system quality, and lack of system testing.
- Project definition – inadequate goal definition, wrong budget evaluation, bad needs analysis, and lack of a good business case.
- Employees – negative attitudes of employees towards changes, lack of adequate knowledge and skills needed for the ES adoption and use, fear for change, and employees' habits.
- Preparation – issues related to bad preparation of implementation project, such as constraints resulting from public procurement law, inadequate pre-implementation analysis, and wrong decisions during the preparatory stage of the project.
- Change management – difficulties with change management process mainly referring to problems with communication between the project team and other employees within the company and poor quality training of employees.
- Top management – issues related to lack of top management support, their inadequate involvement and lack of leadership.
- Finance – financial factors mainly related to unexpected expenses during the implementation project and lack of financial resources.
- Provider – issues referring to lack of good cooperation between the provider and adopter companies and lack of support from the provider.
- Project status – factors describing low status of ES implementation project, such as perceiving ES adoption as an IT project with only IT department responsible for the project and low priority of ES implementation tasks.
- Project management – issues related to bad project management and lack of adequate coordination of activities within the project.
- Schedule – time-related issues such as lack of time for ES implementation and tolerating delays in meeting deadlines.

- Project team – factors describing lack of competence of people recruited for the project team, i.e. project manager and project team members.
- Infrastructure – inadequate hardware and network infrastructure for the new enterprise system.
- Business process change – issues related to organizational changes: some respondents point at lack of BPR as a failure factor; however, on the other hand, other respondents illustrate negative consequences of too radical BPR.

4.2   Causes of Failure by Respondent Age

The results of data analysis from the employee age perspective are presented in Table 1. The table displays the distribution of reported failure factors across the three age groups. The bullets in the table were defined on the basis of the perceived importance of factors provided by the respondents from an individual age group.

Table 1. Failure factors by age

Failure factor	Younger	Middle-aged	Older
System	●	◐	●
Project definition	●	◐	
Employees	◕	◔	◐
Preparation	◔	◐	◐
Change management	◐	◔	◐
Top management	◔		●
Finance	◐		◐
Provider	◐	◔	
Project status	◔		◐
Project management	◔	◐	
Schedule	◐	◔	
Project team	◔	◔	
Infrastructure	◔		
Business process change	◔		

Note: ● – very high, ◕ – high, ◐ – medium, ◔ – low level of importance

The most important reason of ES implementation failure is related to problems with the system. This reason is perceived regardless of respondent age. However, the way of perception of this problem appears to change with age. In particular, perceptions of system-related issues associated with system choice and lack of fit (which refers to the current company situation) are growing with age. However, the youngest respondents' observations refer to threats related to detailed technical issues such as data import, system efficiency, and testing.

The second most important failure factor, perceived regardless of respondent age, refers to problems with employees. In this context, younger respondents point at employees' knowledge and experience, while older workers emphasize the role of employees' attitudes. The next two failure factors reported by the respondents from each age group are project preparation and change management. In the case of project preparation older respondents were more specific in their opinions and pointed at detailed issues (e.g. legal regulations within an organization), while the youngest interviewees described the related failure factors more vaguely (e.g. bad preparation for the implementation project, inadequate pre-implementation analysis). Nevertheless, in the case of change management-related issues the youngest respondents perceived a greater spectrum of failure factors, while older interviewees reported just the issues associated with communication.

The remaining causes of failure (10 out of 14 elements) were reported by respondents belonging to at the most two age groups.

Perception of failure causes noticeably decreases with age in the case of the following elements: project definition, provider, and schedule. These issues are clearly related to the implementation project and capture various considerations associated with its definition and run.

The oldest respondents clearly perceive a failure factor in lack of top management support and, to a somewhat smaller extent, in financial aspects and project status. In their evaluation of these threats, the oldest workers are accompanied by the youngest employees; however, to a lesser extent. It might be noticed that the emphasized threats are associated with the company's organization and condition and the company's perception of the implementation project.

The oldest respondents do not report whatsoever such issues as project definition, project management, schedule, and project team. These issues are generally associated with the project definition and run. Only the last element, project team, relates to people; however, in the context of project organization.

Causes of failure related to infrastructure and business process reengineering are to a small extent reported merely by the youngest respondents. It can be emphasized here that the youngest interviewees describe these failure factors in a more detailed way, which might indicate their considerable involvement and experience in this field.

5 Discussion and Implications

5.1 The Most Important Failure Factors by Age Group

The oldest respondents perceive threats mainly associated with the company where they are employed. They perceive the implementation project and technology through the perspective of the company. For instance, the important elements perceived by the oldest workers, such as preparation, finance, and project status, are associated with the implementation project; however, they also relate to the company organization. On the other hand, such issues as project definition, provider, project management, schedule, and project team refer first and foremost to implementation project and to a much lesser extent to the company. These issues are not emphasized by the oldest respondents.

It appears that such an attitude of older respondents results from their professional experience and a more holistic perception of the problems within their companies (e.g. [12]). Perhaps it might stem from organizational positions held by older workers and ensuing responsibilities. A very strong emphasis on the role of top management might be also explained by habits acquired by employees in the past during the time of centrally planned economy, where companies were highly dependent on top management personnel. In addition, older workers might reveal some shortcomings in modern approach to management, which might explain to a certain extent lack of perception of threats in the area of project management. Differences in the approach to project management with respect to age were also highlighted in prior research (e.g. [19]).

The youngest respondents notice first and foremost threats associated with technology and they appear to perceive the technology-related considerations in the most complete way. The youngest workers emphasize to a similar degree the negative consequences of poor project definition for the implementation process successfulness. The issues associated with people are also perceived by the youngest workers; however, to a somewhat smaller extent compared to the perception of other failure factors. Such prioritization of failure causes by the youngest respondents might be explained by their experience and competence in ICT and by their roles in the implementation projects. A high level of ICT competence is usually required at specialist positions in organizations. The role played in an implementation project and impediments experienced during the project run, perceived from a certain organization position, might have an impact on reported threats (e.g. problems with data migration, low quality and efficiency of the implemented enterprise system). Similar observations with respect to the youngest employees were achieved while analyzing the role of age in the perception of barriers to ES adoptions [21].

The middle-aged respondents perceive to the smallest extent dangers associated with technology, as compared to other age groups. This might be due to the fact that they cope with technology well and they are reasonably well prepared and educated. However, they might reveal a lower level of technology-related awareness than the youngest respondents. Interestingly, the middle-aged respondents perceive issues associated with the implementation project fairly clearly and in this respect they are similar to the youngest respondents. However, they emphasize the role of project management stronger than the youngest respondents. Such a situation might be explained by the duties and responsibilities resulting from the managerial positions often held by the middle-aged respondents during the implementation project. Such positions require substantial experience gained in the focal organization and knowledge about modern management techniques. The specific work ability of workers in the middle age group (35–50) is also suggested by the research on age and productivity at work. In particular, Ericsson and Lehmann [7] argue that it takes roughly 10 years to achieve expert competence in situations where strategic and analytic competences are important. In the same vein, Skirbekk [17] estimates that the peak productivity potential occurs in ages 35–44 years.

5.2   Recommendations

The respondents, regardless of their age, indicate system-related issues as the most important failure factor. Therefore, undoubtedly, in ES implementation projects it is advised to put special emphasis on system choice, its quality, and fit to the company's needs. However, it should be noted that both adopting organization and system provider are responsible for any difficulties in this area. It is therefore important to involve both organizations while defining project requirements and responsibilities. Thus, it appears that shortcomings in project definition and preparation are more source causes of failure than system-related considerations. A necessary condition of good project preparation is related to competent employees and top management involvement. This issue is well illustrated by one respondent, a system provider representative, who stated: "It often happens that companies want changes, but they do not know how these changes should look like. They frequently change requirements during the implementation project. Often, at the client companies, there are no competent people who posses both IT knowledge and are empowered to proceed with the agreed upon regulations".

The considerations of ES adoptions indicated as the most important failure factors (i.e. system quality and fit, project preparation and definition, and employees involved in the project) are related with each other. It appears that potential of employees at various age should be properly used in order to overcome those threats. Gargeya and Brady [9] pointed at inadequate internal readiness and inappropriate planning as the paramount causes of ES project failure. They emphasized the role of managers in understanding of the implications of the system, making decision about the changes, and disseminating the information about the changes to their subordinates. It appears that incorporating the perspective of employee age might help in gaining deeper insight into the importance of internal readiness and inappropriate planning. Our research clearly indicated that older employees (i.e. middle-aged and the oldest) emphasize to a greater extent the significance of employees' negative attitudes and poor project preparation as factors contributing to the project failures. On the other hand, the youngest respondents point at shortcomings in employees' competence and consequences of poor project definition. The current study's findings therefore suggest that good cooperation between employees at various age should contribute to better preventing the occurrence of standard causes of ES adoption failures.

A recommendation for practitioners, resulting from our investigation, is a suggestion to staff project teams taking into consideration employees' age and project stage. It is advised to create a team consisting of empowered, knowledgeable, and experienced people from both adopter and provider side in order to establish the key project guidelines. Middle-aged and older employees are especially recommended for this role. However, transferring the strategic guidelines into more detailed system-related goals definitely requires involvement of the youngest employees as experts in new technologies. They should familiarize themselves with the new enterprise system and indicate its pros and cons as compared to legacy systems. Naturally, in order to obtain a proper choice, a good practice would be to perform such a procedure for several system providers.

Another suggestion refers to taking into consideration the employees' age structure in the company at present and in the future, which is important due to an expected process of workforce ageing. In this respect, it is necessary to create an adequate system interface in order to help older employees to get accustomed to changes (e.g. ample fonts size, simple design, avoiding unnecessary applications, adequate colors). It is also advised to plan a sufficient time for trainings and involve adequate consultants from the provider company. It is recommended to include these considerations in the agreement with the implementation services provider.

It appears that a useful solution is to secure cooperation between the youngest employees and their older workmates in order to make use of strong points of these two groups. Such cooperation might be useful in working out an optimal system solution: the youngest might elaborate detailed system requirements, which are next tested by older employees. In addition, the younger can learn accuracy in work and loyalty to the company from older workers. Older employees, in turn, might supplement their IT knowledge and overcome fear towards changes and new technologies (e.g. [12]).

5.3   Limitations and Future Research

The main limitation of the current study is associated with the fact that the research was based on the data gathered in one country, i.e. Poland. Therefore, the scope of research findings refers first and foremost to Polish practitioners. Poland is perceived as a transition economy, i.e. an economy being in transition from a centrally planned system to a free market system [16]. As suggested by prior research, such countries might experience different ICT-related considerations than well-developed economies (e.g. [24]). In consequence, the application of the current study's findings to more developed economic settings should be done with caution. Nevertheless, as similar changes in workforce are taking place in other EU countries [26], the current study's findings might be applied, to a certain extent, to other countries within the European Union.

The current study's limitations suggest that one of the interesting avenues of future research might be an in-depth comparison of age-related considerations between transition and well-developed economies. Such an investigation would yield interesting insights into the role of a country's economic development in ICT adoption factors from an employee age perspective. Future research might also focus on functions served in the implementation project and organizational positions within companies held by representatives of individual age groups.

6   Conclusion

The current study examined the role of employee age in the perception of factors contributing to failures of enterprise system implementation projects. The study builds on the experience of ES practitioners from Poland. Using data-driven approach, the discovered elements were divided into fourteen failure factors, among which the most important issues are related to system, project definition, employees, company's preparation, change management, and company's top management. In order to investigate the role of respondent age in failure factor perception, we divided the respondents into the following age groups: the young, the middle-aged, and the older. The results of the age-based analysis suggest that perceived ES failure factors are related to employees' age. In general, older employees tend to perceive ES failures from the perspective of a company and its organization, while younger workers recognize causes of ES failures mainly in the organization of implementation project. The current study's findings suggest that potential of employees at various age should be properly used during team building at different project stages to better preventing the causes of ES adoption failures.

Acknowledgments. This research has been financed by the funds granted to the Faculty of Management, Cracow University of Economics, Poland, within the subsidy for maintaining research potential.

References

1. Amid, A., Moalagh, M., Ravasan, A.Z.: Identification and classification of ERP critical failure factors in Iranian industries. Inf. Syst. **37**, 227–237 (2012)
2. Boersch-Supan, A.: The impact of global ageing on labour, product and capital markets. In: Cabrera, M., Malanowski, N. (eds.) Information and Communication Technologies for Active Ageing, pp. 7–34. IOS Press, Amsterdam (2008)
3. Charmaz, K.: Constructing Grounded Theory. A Practical Guide Through Qualitative Analysis. Sage, London (2006)
4. Chen, C.C., Law, C.C.H., Yang, S.C.: Managing ERP implementation failure: a project management perspective. IEEE Trans. Eng. Manag. **56**(1), 157–170 (2009)
5. Conen, W., van Dalen, H., Henkens, K., Schippers, J.J.: Activating Senior Potential in Ageing Europe: an Employers' Perspective. Netherlands Interdisciplinary Demographic Institute (NIDI), The Hague (2011)
6. Corbin, J., Strauss, A.: Grounded theory research procedures, canons, and evaluative criteria. Qual. Sociol. **13**(1), 3–21 (1990)
7. Ericsson, K.A., Lehmann, A.C.: Expert and exceptional performance: evidence of maximal adaption to task constraints. Annu. Rev. Psychol. **47**, 273–305 (1996)
8. European Commission: The 2015 Ageing Report. Economic and budgetary projections for the 28 EU Member States (2013–2060). European Economy 3 (2015)
9. Gargeya, V.B., Brady, C.: Success and failure factors of adopting SAP in ERP system implementation. Bus. Process Manag. J. **11**(5), 501–516 (2005)
10. Glaser, B., Strauss, A.L.: Discovery of Grounded Theory: Strategies for Qualitative Research. Aldine, Chicago (1967)
11. Hawari, A., Heeks, R.: Explaining ERP failure in a developing country: a Jordanian case study. J. Enterp. Inf. Manag. **23**(2), 135–160 (2010)
12. Ilmarinen, J.: Ageing workers. Occup. Environ. Med. **58**, 546–552 (2001)
13. Lech, P.: Causes and remedies for the dominant risk factors in enterprise system implementation projects: the consultants' perspective. SpringerPlus **5**(1), 238 (2016)
14. McMorrow, K., Roeger, W.: The Economic and Financial Market Consequences of Global Ageing. Springer, Heidelberg (2004). doi:10.1007/978-3-540-24821-7
15. Prskawetz, A., Fent, T., Guest, R.: Workforce aging and labor productivity: the role of supply and demand for labor in G7 countries. Popul. Dev. Rev. **34**, 298–323 (2008)
16. Roztocki, N., Weistroffer, H.R.: Information technology investments in emerging economies. Inf. Technol. Dev. **14**(1), 1–10 (2008)
17. Skirbekk, V.: Age and productivity potential: A new approach based on ability levels and industry-wide task demand. Popul. Dev. Rev. **34**, 191–207 (2008)
18. Soh, C., Chua, C.E.H., Singh, H.: Managing diverse stakeholders in enterprise systems projects: a control portfolio approach. J. Inf. Technol. **26**(10), 16–31 (2011)

19. Soja, E., Paliwoda-Pękosz, G., Soja, P.: Perception of difficulties during enterprise system adoption and use: the role of employees age. In: Proceedings of the 10th International Conference Accounting and Management Information Systems AMIS 2015, Bucharest, 10–11 June, pp. 50–65. The Bucharest University of Economic Studies (2015)
20. Soja, E., Soja, P.: Exploring the role of employee age in improving ICT adoption projects: lessons learned from enterprise system practitioners. In: Proceedings of the 22nd Americas Conference on Information Systems AMCIS 2016, San Diego, 11–14 August, pp. 1–10. Association for Information Systems (2016)
21. Soja, E., Soja, P.: Towards Understanding the Role of Employee Age in ICT Adoption: Learning from Barrier Perceptions of Enterprise System Practitioners. In: Proceedings of European, Mediterranean & Middle Eastern Conference on Information Systems EMCIS 2016, Krakow,, 23–24 June, pp. 1–13. Cracow University of Economics (2016)
22. Soja, E., Soja, P., Paliwoda-Pękosz, G.: Solving problems during an enterprise system adoption: does employees' age matter? In: Wrycza, S. (ed.) SIGSAND/PLAIS 2016. LNBIP, vol. 264, pp. 131–143. Springer, Cham (2016). doi:10.1007/978-3-319-46642-2_9
23. Soja, E., Stonawski, M.: Zmiany demograficzne a starsi pracownicy w Polsce z perspektywy podmiotów gospodarczych. In: Kurkiewicz, J. (ed.) Demograficzne uwarunkowania I wybrane społeczno-ekonomiczne konsekwencje starzenia się ludności w krajach europejskich, pp. 173–210. Wydawnictwo Uniwersytetu Ekonomicznego w Krakowie, Kraków (2012)
24. Soja, P., Cunha, P.R.: ICT in transition economies: narrowing the research gap to developed countries. Inf. Technol. Dev. **21**(3), 323–329 (2015)
25. Umble, E.J., Umble M.M.: Avoiding ERP implementation failure. Industrial Management (2002)
26. United Nations: World Population Prospects: The 2015 Revision. Department of Economic and Social Affairs, Population Division (2015)
27. Van Dalen, H.P., Henkens, K., Schippers, J.: Productivity of older workers: perceptions of employers and employees. Popul. Dev. Rev. **36**(3), 309–330 (2010)
28. Volkoff, O., Strong, D.M., Elmes, M.: Understanding enterprise systems-enabled integration. Eur. J. Inf. Syst. **14**, 110–120 (2005)
29. Wagner, N., Hassanein, K., Head, M.: Computer use by older adults: a multi-disciplinary review. Comput. Hum. Behav. **26**, 870–882 (2010)
30. Wong, A., Scarbrough, H., Chau, P.Y.K, Davison, R.: Critical failure factors in ERP implementation. In: Proceedings of Pacific Asia Conference in Information Systems (2005)
31. Zhang, Z., Lee, M.K.O., Huang, P., Zhang, L., Huang, X.: A framework of ERP systems implementation success in China: an empirical study. Int. J. Prod. Econ. **98**(1), 56–80 (2005)

A Proposal for an ECM Systems Modeling Method – Defining Tactical Perspectives - Lesson Learnt from a Case Study

Jan Trąbka[✉]

Department of Computer Science,
Cracow University of Economics, Cracow, Poland
Jan.Trabka@uek.krakow.pl

Abstract. Concepts, technologies and tools gathered together under the common label of Enterprise Content Management (ECM) are currently the most dynamically developing area of Management Information Systems. The author's literature research and practical experience show that there are no sufficient operational analysis and design methods dedicated to ECM. The main focus of this paper is the case study of a big multi-branch organization from the sector of medical services which the author has used to make an attempt at showing tactical perspective and guidelines for creating an ECM implementation model of the enterprise. The described set of integrated perspectives includes: Content, Processes, Organizational Structure, Localization Structure, Business Rules and IT-environment. That set is the foundation of the author's proposal of comprehensive analytical method named Enterprise Content Management Modeling Method (ECM3). This paper is the first stage of the author's presentation of the ECM3.

Keywords: Enterprise Content Management (ECM) · Unified Content Strategy · Workflow · BPM

1 Introduction

IDC analysts [2] in their report "The Digital Universe in 2020" estimate that the amount of data managed worldwide in the years 2005–2020 is going to rise 300 times, from 130 EB (1 million TB) to 40 000 EB. The report's authors underline the fact that the increase is in 80% generated by enterprises. To answer the question of what kind of data will be most often dealt with by enterprises it is worth quoting Mancini, who believes that 90% of data managed are unstructured, i.e. "*it is data that cannot be put into rows and columns*" [6]. Unstructured data do not only include documents of all kind but also information exchanged in emails or social media as well as multimedia, i.e. photos, films or sound records. The growing and strategic rise in the meaning of the content concept embracing both structured and unstructured data should not be surprising. These issues come together within the domain of Enterprise Content Management (ECM) and merge into concepts, technologies and tools. The following description has been adopted for the definition of the key content concept: it is a collection of structured and unstructured data, information and knowledge available in

© Springer International Publishing AG 2017
S. Wrycza and J. Maślankowski (Eds.): SIGSAND/PLAIS 2017, LNBIP 300, pp. 136–151, 2017.
DOI: 10.1007/978-3-319-66996-0_10

the electronic format (e.g. database records, digitalized documents, electronic documents, emails, messages sent through social media or sound and image recordings) as well as the traditional format (i.e. paper or microfilm) [14]. ECM treats enterprise processes which do not concentrate on the resources such as products, materials, or machines (characteristic of ERP systems) but processes connected with creating, storing and distributing content.

The characteristics and at the same time main challenges connected with ECM are the following:

- unstructured format of the processed content and the diversity of media used to distribute it,
- a huge and constantly extended set of technologies (devices and software) used for content processing,
- access to data processing for all of the enterprise's employees,
- content processing often occurs ad hoc (conceptual, group work),
- content processing in an enterprise requires close integration with fundamental data sets of employees, organizational units, clients, etc. At the same time a considerable number of ECM technologies need external interfaces compatible with their business partners' or clients' systems.

Simons and vom Brocke in their deliberation on the current position of ECM concept within the scientific field of Information Systems (IS) stress that due to the broad and strategic, both organizational and technological, scope of problems the ECM concept contains, it is an important research subject in the IS discipline. Still the above-mentioned authors while analyzing the current state of the research on the ECM concept conclude that it "*has been largely ignored by the IS discipline and can be characterized as bereft of theory*" [11].

This observation is confirmed by the author's literature research and above all practical experience gained from projects of building an organization implementing ECM models. The research allows to state that currently there are no complex analytical methods dedicated to ECM systems and the methods known from theory and practice do not embrace all of the above-mentioned characteristics and challenges. This paper presents a thesis that the concept of content and new information technologies for its management that are integrated in the ECM concept, introduce totally new requirements for methods as well as tools that have been used to create the so far existing enterprise information systems. The author's long-term research aim is to propose a complex method for modeling an organization preparing itself for an ECM class system implementation. This paper is the first stage of the process. The research aim of this paper is to show tactical perspectives of seeing the enterprise implementing ECM. These perspectives will be the foundation on which tools and processes of the newly created method can be built on. On the basis of his own analytical work in an ECM platform implementation project run in a big Polish medical enterprise, the author described a set of tactical and integrated perspectives supposed to form a framework for an organization preparing itself for the ECM implementation model. The method has been operationally named the Enterprise Content Management Modeling Method (ECM3). In order to discuss the paper's assumptions the author begins with a literature overview presenting the sources used to build the author's new method (Sect. 2).

Section 3 discusses the assumptions of the case study presented in the paper. The main part of the paper (Sect. 4) discusses pre-implementation analysis in the studied case and shows the explored perspectives of seeing the ECM model. This section is concerned with the place and meaning of each of the perspectives in the analyzed case. The next Sect. 5 presents the discussion and the author's proposition for the ECM3 method's tactical perspectives. Section 6 presents limitations and defines further steps to be taken in order to achieve the main research aim – building a complex method of ECM systems modeling. The last section presents conclusions of the paper.

2   ECM3 Method Sources - Literature Review

2.1   Enterprise Content Management – Its Concept and Main Technologies

The concept of ECM was introduced in 2001 by the Association for Information and Image Management (AIIM). The definition of the ECM concept proposed by AIIM at that time over the years has been frequently extended. One of the most quoted definitions was proposed by Smith and McKeen. According to them it is *"the strategies, tools, processes and skills an organization needs to manage all its information assets (regardless of type) over their lifecycle"* [12, p. 648]. The authors pointed to the integrated character of the ECM concept, stressing that it goes beyond single applications, business areas, processes or functions to grasp all of the organization's information resources irrespective of their type, format, specificity and source. Päivärinta and Munkvold put emphasis on the integration and strategic scope of the ECM concept and called it *"a modern, integrative perspective in information management"* [7, p. 96].

However, ECM is not an original concept in terms of technology. Its constituent technologies were used much earlier and included Document Management and Content Management [13]. Vom Brocke and his team broadened this set with Web Content Management and Records Management [17]. As ECM systems are not only content repositories but equally importantly they ensure the dynamic processing of content and hence workflow technology and BPM (currently known as Content Workflow) have also been added to the list above. Gartner analysts expanded the above-mentioned set with Image-processing applications and Social Content [3]. A closer specification of the aforementioned technologies can be found in [14]. Figure 1 presents ECM's main technological components. The diagram is supplemented with functional weights assigned to particular technologies by Gartner's analysts in the most recent "Magic Quadrant for ECM 2015" report. Functional weights are understood *"as belonging to the core functionalities that an ECM platform should be able to fulfill"* [5].

The set of technologies presented clearly indicates that ECM's main objective is to manage unstructured data. Smith and McKeen point to the fact that currently the main challenge organizations face is to decide which unstructured data will be managed and how this will be performed, and which data will remain unstructured thus unmanaged [12]. Smith and O'Callaghan draw attention to the fact ECM is not only a technological but above all an organizational phenomenon. From an organization's perspective the

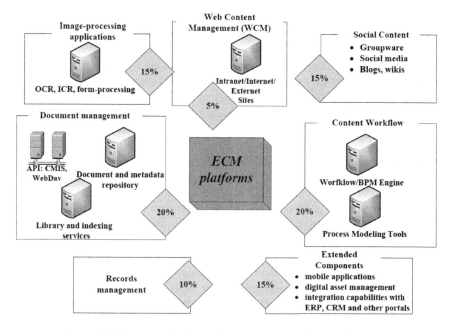

Fig. 1. ECM main technological components and functional weights

principle objective is to analyze the information and content needs reported by the organization's units and to decide in what way they are going to be resolved [13].

The set of ECM's technological components presented above affects a wide range of uses of this class of systems in today's enterprises. The enterprise's functional areas most often supported with ECM systems were indicated in [14]: managing office documents circulation with a central document repository, project management, quality processes management (including ISO), budgeting, financial documents flow, staff and payment processes, IT services management or intranet sites management. Scott describes the use of ECM platform as an elaborate knowledge management system for 10 000 employees and clients of a global corporation [10]. Rockley [9] when presenting her strategy of content management mentions it can be applied to customer relationship management (CRM) and handling publishing processes or medical documentation history. These examples prove that the wide and universal range of ECM technology provides unlimited and constantly broadening spectrum of applications in present-day enterprises. Due to the component structure and versatility of the ECM systems they will be referred to as ECM platforms in the further part of this paper.

2.2 Strategic Approaches to Content Management

The discussion of strategic approach to content management is methodologically based on "A framework for ECM research" [16]. The framework proposed consists of four strategic and integrated perspectives: content, technology, processes and enterprise. Content perspective concerns identification of content items, their semantics, structuring, and organization. Technology perspective encompasses the development of hardware, software, and standards for content management in organizational context. Process perspective considers the development and deployment of new content management solutions in enterprises. The enterprise perspective considers organizational, social, legal, and business issues of content management. "The Unified Content Strategy" (UCS) [9] or the "ECM - blueprinting framework" [18] may serve here as good examples of detailed strategies contained in the framework of the defined perspectives, described in literature.

The first one, undoubtedly most popular in practice and most often quoted strategy is aimed at creating intelligent frequently reusable content presented on many diverse system and device platforms. It is dedicated to organizations creating and distributing substantial number of documents which upon transformation according to UCS will be called information products (IP) (these may be ad leaflets, product cards, brochures, instructions, etc.). UCS is founded on the intelligent and adaptive content concept. Intelligent content is content that is structurally rich and semantically categorized, and is therefore automatically discoverable, reusable, reconfigurable and adaptable [9]. Creating intelligent content starts with the modeling stage. Content Modeling is a process of defining the content's structure and granularity (particularity). Information product model represents the hierarchical order of the components. Component is a block of content of a defined semantic type. Single semantically defined components may be used for various IP, e.g. the "product description" component will function in all, a brochure, a product card or a leaflet. By defining templates responsible for their display on web sites or mobile devices we are able to maintain topicality of the content and radically lower the cost of its processing. UCS enables extracting content from documents and exchanging the concept of document management with the concept of content management.

Approach "ECM - blueprinting framework", mentioned above focuses on incorporating processes connected with the content's lifecycle into operational processes of an organization. The key feature of the approach is the application of the so called ECM-blueprints, referential models of handling processes of content consistent with content lifecycle activities (e.g. capturing, creating, editing, or archiving content) [18]. The approach discussed undoubtedly provides a faster and cheaper implementation of content processes. At the same time, the authors stress that it is a much more strategic approach which requires developing solutions on tactical and operational levels.

A review of research frameworks and selected strategies provides a general high-level perspective on an organization implementing ECM. A literature review shows what a challenge it is to implement ECM class systems and points to the need of developing complex, tactical and operational methods of their analysis, design and implementation.

3 Research Methodology

3.1 Definition of Case Study

To discover and build (and next verify) the ECM3 method the author employed a quality research approach, mainly qualitative case study methodology which provides tools to research complex phenomena in their full environmental context [1, p. 544]. A hallmark of case study research is the use of multiple data sources, a strategy which also enhances data credibility [8]. Potential data sources may include, but are not limited to: documentation, archival records, interviews, physical artifacts, direct observations, and participant-observation [1]. The presented case study belongs to the exploratory type [20].

The subject of the case study presented in the paper is a big Polish company from the medical services sector engaged in the process of choosing and implementing an integrated ECM platform in a number of its key operational areas. The company in question agreed to publishing of this paper on the condition its name is not disclosed. The choice of this particular company for the case study's subject was dictated by a few factors:

- the company's scale of operations and multi-departmental structure,
- wide scope of the ECM implementation project: handling the entire incoming and outgoing documentation including cases management, financial documentation flow and handling of documentation and quality processes (consistent with a few norms). Each of the areas listed to different extent utilizes various ECM technologies, however, from the perspective of the whole project it covers the whole structure of ECM components (Fig. 1),
- The company was not considering handling particular areas with parallel applications but preferred one integrated platform for all of them. What is more the areas included in the project were not supported by uniform IT systems. Thus, we are dealing with a real modeling, designing and implementing a system from scratch.

3.2 Projects Background

The organization analyzed in the paper is the one of the biggest Polish network of medical diagnostic laboratories. The network consists of 140 laboratories (and 400 collection stations). The network performs 28 million tests a year and serves 10 million patients. The enterprise employs about 4 000 workers. The places where the company operates are proportionally located across the territory of Poland. Such a complex physical location structure as well as the organization's management style have led to developing a division of the organizational structure into the Headquarters and 8 Regions (which cover the whole country). The managerial and supportive functions for the whole organization are centered in the company's headquarters divided into Central Departments (Financial, HR, IT, etc.). The company has a certified quality management system ISO 9001 as well as branch certificates PN-EN ISO 15189:2008 and PN-EN ISO/IEC 17025:2005 and PN-ISO/IEC 27001:2014-12.

To demonstrate the extent of the company's operations in the context of an ECM platform implementation it is worth referring to the following statistics. The documents circulation over the period of one year counts over 200 000 documents. The quality management system's repository holds 150 000 documents.

The structural and territorial development of the company and the amount of information especially in the form of documents, gave rise to the Board's decision to start a project of selection and implementation of an ECM. The project was divided into a few areas dealing with: the incoming and outgoing correspondence, so called office documents circulation, financial documents circulation and acceptance (purchase and business trip invoices, etc.) and quality management system documentation (with the required creation, acceptance and distribution processes). The author of the paper played the role of an external expert responsible for creating the work schedule and analytical documentation's framework, training the team for the tools used and the whole process's coordination and verification. The project started with a pre-implementation analysis conducted by the company itself with the assistance of external experts. The stage was supposed to prepare the project's assumptions and specifications for the most important processes in the above-mentioned areas. All stages of the project were planned for two years and at the moment it is still being realized.

4   Discovering Specific Perspectives of ECM Modeling – Practical Lesson Learnt

At the beginning of the pre-implementation analysis a classic structural approach was adopted which recognizes two basic perspectives on information system description – processes and data [21]. Structural approach tools are less popular, however, the above-mentioned strategy proves true for many more recent integrative methods like ARIS or BPMS (BOC companies). The perspectives can be also found in Enterprise Architecture building approaches. Content and processes perspectives are also mentioned in all of the strategic approaches to content management presented earlier in the paper (see Sect. 2.2).

4.1   Content Perspective

Content replaces the data perspective. Yet one should state that the latter one was understood as unstructured, most often relational data. The content perspective embraces all structured, semi-structured and unstructured data. All of these appeared in the project analyzed during the content modeling process. One could also note correlations between the content types and the functional areas modeled.

A quality management system is a typical example where unstructured content is extensively created and distributed inside an organization. We are speaking here about the quality management documentation including quality manuals, procedures, instructions and many more. The organization described in the paper has a number of normalized management systems (see Sect. 3), each of them provided with its own set of documentation. It should not surprise that the repository contains about 150 000 documents. In such cases the basic requirement from an ECM platform to be satisfied is to provide a central and open for general use document repository supporting the processes of document preparation, distribution and audits. Analytical work presented in the paper was performed using the Unified Content Strategy (UCS) [9]. The strategy consists in creating intelligent, free-format, adaptive and reusable content (see Sect. 2.2).

This stage of the project included content modeling. Various kinds of quality documents modeled were given attributes (metadata) with reference to their categorization, identification or so called library attributes, but most importantly attributes modeled the documents' semantic structure. The information product models (named so according to UCS to differentiate them from the document concept) were presented using XML schemas. The information products alone are also stored as XML documents. Below presented an example of a medical procedure. The XML tags shown in the example below serve to display the semantic structure of the document.

```
<Medical Procedure >
    <Header Number="IB/LAB/913" Author="John M." Title="CAE levels
    measurement with COBAS 600" Version="I" Entry_date="2012-02-06"/>
    <Objective_of_the_manual>The manual describes how to  perform a
    quantitative measure of carcinoembryonic antigen  concentration in
    blood serum </Objective_of_the_manual >
    <Manual_description>
      <Test_requirements>
        <Reagents>R1 biotin-labeled anti-CEA antibodies an 18 ml
        container</Reagents>
        <Calibrators> CEA CalSet II </Calibrators>
      </Test_requirements>
      <Procedure>
        <Step Name="Specimen preparation" Parameter stability in the
        whole blood - 20-25 C - 7 days Specimen type: blood serum S
        pecimen stability: 7 days at 2-8 °C </Step Name>
```

[Medical procedure as information product. Application of the intelligent content concept]

It is worth noting that in this case we are dealing with a total unstructured to semi-structured content transformation.

One more kind of transformation was identified in the financial circulation area which requires both the circulation of the document's electronic source version and its subject matter presented in the form of attributes. This applies to page identifiers, descriptions and transaction values as well as other supply parameters. These attributes are extracted (mainly manually with the OCR support) from financial documents during the registration, description or acceptance steps. The final set of attributes has to be sufficient to create (often automatically) an entry in the financial and accounting system. The creation of financial document models began with the recognition of relative data structure in the target ERP system databases. This kind of transformation changes unstructured content into structured content.

As far as analytical tools are concerned one can state that in the context of structured data there are many described tools. The unstructured content area does not have any standards developed. The subject of content modeling tools is going to be the next element of the ECM3 method and will be described by the author in his next articles.

The ECM platform map has been identified as another sub-perspective of the content perspective subject to analysis. Currently ECM platforms, especially their repositories become the central storing place for an organization's content. This stems from the fact content processes concern all of the organization's employees. In this case study each of the employees, with no exception, had to be given an access to all of the main areas mentioned. The need for the platform's universality exposed two issues. First, different organizational units want to have their own place on the platform, especially in the repository where a unit can store public (available to other departments) and private content for internal use (it can also realize group work processes presented in the next sub-section). One should remember that ECM technologies include Web Content Management components (a derivative of CMS technology). Platforms enable a logical division into sites. A site has its own presentation pages, own part of the repository, users and permissions systems and many functionalities available: information walls, discussion lists, calendars etc. This makes an ECM platform not only a central content depository but also the primary internal communication tool in an organization. The second issue identified, deriving from the first one, is how to build the platform's logical structure and make it easily navigable. The ECM platform map can be the solution here. The map contains the sites structure divided into functions (office circulation, business trips, quality management) and organizational units (HR, medical technologies department, etc.). A map of the platform's content is a perspective dedicated to ECM systems and does not have any established modeling tools. While realizing the project a few of the author's own proposals were put forward, which will be presented in the ECM3 method's tool section.

The IT environment, tightly connected with the content perspective, is the next perspective to be discussed.

4.2   IT Environment Perspective

When modeling various kinds of content one should consider their sources and places of distribution. The sources may include the enterprise's internal systems, or the external systems of its contractors and partners. This sub-perspective has been named **Interfaces**. The next IT modeling aspect is the place and requirements of an ECM platform in an organization's **System and Hardware Environment**. The two sub-perspectives due to their mutual relations and the fact that they complement each other have been collectively called the **IT environment**.

The first issue that appeared when modeling the internal interfaces perspective was how to obtain and update the data of the platform users. Dealing with 4 000 employees/users it is not possible to manage them independently without any central administrative systems. One had to determine the principles of an "ECM platform user's system journey," i.e. the period from the first day to the last day of the user's employment. The user's data are collected from domain controllers (LDAP), mailing system, and HR system. Upon collecting users' data by the mechanisms inside the platform appropriate permissions are assigned to the employee's position in the organization's structure (the structure has to be modeled in ECM – this aspect will be described in the following sub-sections).

The contractors' data were the next key area of the basic data identified in the project. The data were sourced from three systems: ERP, CRM and domain system supporting essential laboratory activities. Due to handling the office document circulation the ECM platform manages the largest collection of contractors as regards the data range and number. The collection is the sum of the above-mentioned systems contents. The project included modeling particular interfaces (most of them were bidirectional) and in some cases even the data flow paths were changed. Once the map of internal interfaces was completed it turned out that the ECM platform was going to connect with all of the organization's most important systems (the above-mentioned list extended with BI).

As far as external interfaces are concerned one should ask what medium will be used to receive and send documents to external subjects. The question concerns both the office and financial areas. The financial area distinguishes three forms of documents delivery (invoices and other documents regarding costs): paper documents sent by mail, document images as graphic files (mainly pdfs) sent by e-mail, electronic documents sent in agreed on EDI/XML formats. Each of these channels required modeling a separate interface and taking into account alternative paths in processes models. The interfaces are also bidirectional.

The office processes, especially paper documents received by the company are affected by the system and hardware environment perspective. In order to provide efficient scanning of thousands of documents received by different organizational units over the territory of the whole country it was necessary to model so called places of document scanning and equip them with scanners of parameters proportionate to the estimated number of documents circulating. Each of the scanning places has to be equipped with barcodes printers and readers. This is closely connected to the localization structure perspective to be described in the next sub-section.

4.3   Processes Perspective

In the project two main kinds of processes were identified: ordered workflow processes and ad hoc group work processes. Ad hoc means the processes do not have univocally defined sequence and scheduled completion dates. Workflow processes and group work sub-perspectives were introduced for the use of the ECM3 method.

The office and financial area works started with modeling the processes of documentation circulation as specified by the existing directives and so called office instructions. We are dealing here with typical workflow processes which are directed in line with the defined organizational rules. Since the processes models were created in the "should be" scenario, activities connected to scanning, text recognition (OCR) or barcodes operations were introduced instantly. We can well use here the recommendations of the "ECM - blueprinting framework" approach (see Sect. 2.2). Roles in the processes come straight from the organizational structure. The process engine is able to select the executers of particular tasks independently. The workflow processes in the project were modeled using the BPMN standard.

The quality management processes connected to quality documents distribution and usage approval are also ordered. The processes of group document preparation are ad-hoc in nature. Groups comprising of subject matter specialists, quality management specialists or managers create, consult and express their opinions about particular document versions. Group work processes are based on parallel (sometimes concurrent) or sequential work where successive participants are selected manually at the moment of setting in motion next steps. This type of processes was very troublesome to model. BPMN used in the project to depict such multiple and non-sequential courses led to extremely complicated and most often incoherent models. This subject is extensively discussed within the Case Management term [4]. In practice this sort of processes is sometimes realized without the process engine, directly using the functionalities of concurrent work on documents. Such functionalities are offered by productivity software like MS Office or cloud services like Google Docs.

Two more perspectives, extremely important in the context of content management, appeared during processes modeling: the perspectives of organizational structure and localization structure (name suggested by the author). Both poorly documented in the subject literature.

4.4 Organizational Structure Perspective

During processes modeling analytical teams in each of the functional areas very often referred to organizational structure. It provides answers to the following questions: in what organizational unit and on what position does the employee/user work, who is his/her direct superior, what roles in processes can he/she be assigned? Representing an organization's business rules in the ECM mechanisms would not be possible without copying the multi-leveled structure of organizational units and the dependencies between them. In the case study the organizational structure is tightly correlated with the management style, which is distinctly decentralized, and puts emphasis on the regional structures' decision-making powers. The structure and workflow processes analysis showed that each Region independently goes through most of the defined courses (including booking, which is realized by regional accountants). However, this management rule contains a threat to the processes models since each independent regional structure can "make up" workflow processes paths accordingly to its own experiences and preferences. This would be unusually labor-consuming and the models as well as the subsequent implementation too complex and unstable. These observations and discussions resulted in the Board's decision to unify the processes (universal models) and organizational structures of the whole organization (a regional structure model). Modeling and later on implementing the unified standards made it possible to fully control the workflow processes with a process engine accessing the sets of employees, organizational units (with their relations) and roles. The organizational structure modeling, which in the author's opinion is a very unappreciated perspective within the IS subject, lacks modern notation standards and methods. This was discussed in the author's other paper [15]. The project presented here required the tools described in the above-mentioned article.

4.5 Localization Structure Perspective

A company's organizational structure does not always communicate physical location of its operations. The problem occurs mainly in multi-departmental domestic or global companies. The question to be asked here is which organizational units and employees work at particular physical locations. Bearing in mind ECM it is also important to ask which IT and ICT resources the physical locations have at their disposal. Here we can see the relation to the IT environment. Modeling the network of scanning locations in the project presented took a very long time. To establish them it was necessary to collect all physical locations with information on the number of employees, number of documents, current state of IT and communication resources. The fact that the data collection concerned about 1 000 organizational units (there are 1 400 altogether) in 500 physical locations shows the significance and work-consumption of the perspective. As regards the modeling tools it transpired there were none or they did not meet our the expectations. The used author's own proposals will be described in the tool aspect of the ECM3.

4.6   Business Rules

The business rules perspective is traditionally inseparable from the processes perspective. In the case study workflow processes models were built on the basis of internal directives, regulations and instructions. The influence business rules have on the content is also an important aspect. One can see it very clearly in the quality area where normalized quality systems have compulsory procedures for creating quality documents. The procedures describe formally and in detail the factual range and form of each type of the document required and are the basis for particular content models creation.

The project revealed some correlations. Directives and office instructions were the starting point for office processes modeling. When modeling office processes taking into consideration ECM technology, it becomes apparent that present business rules require updating or creating them all over again. Thus, we started to realize the task of creating and documenting regulations and office instructions observing the ECM technology.

On the most detailed level business rules define the decision-making points in the processes. In this aspect they display a tool choice issue – how to precisely and algorithmically describe a rule so that it could be directly moved to the so called rules engine in a workflow system. This shall be the subject of the tool aspect of the ECM3 method.

5   Discussion – a Proposal for an Integrated Set of Tactical Perspectives of ECM3 Method

The concept of information system creation method was accurately defined by Wrycza [19]. He mentioned a few components where the first two most clearly define the role of conceptual models in the early, analytical and project stages of creating an information system:

- formalism, reality description models – subject domain, its statistics and dynamics, called conceptual models,
- detailed methods and techniques of documenting the system (according to theoretical formalism construction), accompanied by adequate graphical symbolism.

The aim of this paper was to propose an integrated set of perspectives (as particular conceptual models' subjects) which should be considered in order to build a complete model of an organization preparing itself for an ECM system implementation. The practical "lesson" learnt from the case study presented in the previous section and the conclusions drawn from the literature review confirmed that analytical methods used so far do not embrace all of the perspectives required by the ECM systems. Some perspectives are not taken into consideration or are narrowly interpreted. As a result we do not have at our disposal suitable analytical tools and techniques. Preparing the pre-implementation analysis over the last 6 months allowed the author to identify and describe a set of six tactical perspectives which formed the foundations for the ECM3. These include Content, Processes, Organizational Structure, Localization Structure, Business Rules and IT-environment. The meaning and role played by each of the perspectives has been discussed in detail in the previous section and illustrated with real project examples. In the summary we are going to discuss the key characteristic of the proposed set – integrity. Integrity is understood as the mutual relations between the perspectives complementing each other, which constitute a mechanism responsible for controlling the coherence and completeness of the created full model of an organization implementing ECM. The diagram in Fig. 2 presents tactical perspectives of the proposed ECM3 method and their mutual relations.

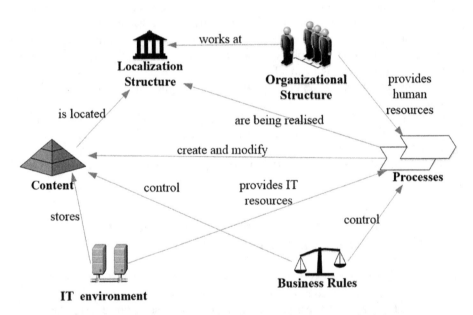

Fig. 2. Tactical perspectives of the ECM3 with their mutual relations

As one can see in the diagram each of the perspectives is linked to at least two other perspectives where the two strategic ECM perspectives of content and processes are indirectly (or directly) connected to all of the others. The relations are complementary, i.e. one perspective cannot be described without being acquainted with the other. The relations are not only conceptual but are also visible on the stage of detailed models creation. They are represented there via common notation elements, e.g. organizational structure connects to processes through the "role" notation symbol, and to localization structure through "localization." The answer to the question of what IT resources are at the organizational unit's disposal may serve here as one more example. In this case one should employ the organizational unit/localization/IT resources relation. One more time it is worth underlining the fact that the relations between the perspectives are a vital mechanism responsible for controlling the coherence and completeness of the full model of an organization in the ECM method.

The presented perspectives are internally divided. Sub-perspectives show a narrower aspect which has its own characteristics and is used differently – this results in the choice of different tools to describe it. For example, the processes perspective was divided into workflow processes and group work processes. One should remember that modeling each of the sub-perspectives requires different tools. For instance, BPMN for modeling ordered workflow processes or CMMN (Case Management Model and Notation) for ad hoc group work processes. The full set of perspectives with their sub-perspectives is presented in Table 1.

Table 1. Perspectives and sub-perspectives of ECM3

Perspective	Sub-perspective
Content	Structured content
	Unstructured content
	ECM platform map
Processes	Workflow processes
	Work group (ad hoc) processes
IT environment	External interfaces
	Internal interfaces
	System and hardware environment
Business rules	Strategic rules
	Operational rules

6   Limitations and Future Research

As it has been mentioned in the introductory section building a set of tactical perspectives of the ECM modeling is the first stage of the ECM3 method's development. The next stage is going to encompass the tools and analytical techniques for respective sub-perspectives. To complete the ECM3 method, the final stage is supposed to comprise a project of analytical procedure – system of stages and tasks, human and infrastructure resources necessary for the method's realization. Selecting the set of CASE tools, which would meet the requirements of the ECM3 will be of particular importance. These stages will be the author's upcoming research subjects.

The set of perspectives proposed in the paper probably is not locked for two reasons. The first reason are limitations connected with the analysis of the enterprise coming from one particular business sector (i.e. medical sector). The scope of the ECM implementation discussed in this case study is really broad and universal (office, financial and quality management documents circulation). Presumably research studies of ECM process in companies from other business sectors will bring new perspectives or sub-perspectives. The second reason are changes in technology and the theoretical knowledge, but even more importantly changes in the practical knowledge of people working in the ECM area. Their practical knowledge is constantly enriched through gaining new hands-on experience. The author of the paper will apply ECM3 in his other projects as it is the best way to verify and improve the method.

7 Conclusions

As it is shown by the reports (quoted in the Introduction) we are witnessing the geometric growth of data processed by enterprises. This growth concerns mainly unstructured data. In this context one should not be surprised by high and constantly growing dynamics of the ECM system market [5]. The literature review and the case study conclusions presented in this paper show that there is shortage of dedicated and ready-to-use modeling methods supporting ECM analysis, design and implementation. Hence, the long-term research aim presented in this paper is to create an integrated modeling method bridging this gap. The paper is the first step in which six perspectives are defined and characterized together with their mutual relations. The described case study became the place for discovering and at the same time verifying their effectiveness. Following those described perspectives comes the challenge of creating operational tools and procedures so that the complete method can be delivered. The author is trying to meet this challenge, but obviously this challenge is valid for the entire ECM domain within the scientific field of Information Systems (IS). Signals coming from the business world show that methods and tools which are yet to be created will be instantly verified and if possible implemented as practical business solutions.

References

1. Baxter, P., Jack, S.: Qualitative case study methodology: study design and implementation for Novice Researchers. Qual. Rep. **13**(4), 544–559 (2008)
2. Gantz, J., Reiznel, D.: The digital universe in 2020: big data, bigger digital shadows, and biggest growth in the far east. In: International Data Corporation (IDC) (2012)
3. Gilbert, M., Shegda, K., Chin, K., Koehler-Kruener, H.: Magic Quadrant for ECM 2014. Gartner Inc. (2014)
4. Hinkelmann, K., Pierfranceschi, A.: Combining process modelling and case modeling. In: Proceedings of the 8th International Conference on Methodologies, Technologies and Tools enabling e-Government MeTTeG14, Udine, pp. 83–95 (2014)
5. Koehler-Kruener, H., Chin, K., Hob, K.: Magic Quadrant for ECM 2015. Gartner Inc. (2015)
6. Mancini, J.: The emperor's new clothes: the current state of information management compliance. AIIM International (2004)
7. Päivärinta, T., Munkvold, B.: Enterprise content management: an integrated perspective on information management. In: Proceedings of the 38th Hawaii International Conference on System Sciences. IEEE, Hawaii (2005)

8. Patton, M.: Qualitative Evaluation and Research Methods, 2nd edn. Sage, Newbury Park (1990)
9. Rockley, A., Cooper, C.: Managing Enterprise Content. A Unified Content Strategy, 2nd edn. New Riders, Berkeley (2012)
10. Scott, J.E.: The knowledge garden and content management at J.D. Edwards. In: vom Brocke, J., Simons, A. (eds.) Enterprise Content Management in Information Systems Research. PI, pp. 183–197. Springer, Heidelberg (2014). doi:10.1007/978-3-642-39715-8_11
11. Simons, A., vom Brocke, J.: Enterprise content management in information systems research. In: vom Brocke, J., Simons, A. (eds.) Enterprise Content Management in Information Systems Research. PI, pp. 3–21. Springer, Heidelberg (2014). doi:10.1007/978-3-642-39715-8_1
12. Smith, H., McKeen, J.: Developments in practice VIII: enterprise content management. Commun. Assoc. Inf. Syst. 11(1), 647–659 (2003)
13. Smits, M., O'Callaghan, R.: Strategy development for enterprise content management. In: vom Brocke, J., Simons, A. (eds.) Enterprise Content Management in Information Systems Research. PI, pp. 91–107. Springer, Heidelberg (2014). doi:10.1007/978-3-642-39715-8_6
14. Trąbka, J.: Enterprise content management platforms: concept update, role in organization and main technologies. In: Pańkowska, M., Palonka, J., Sroka, H.: Ambient Technology and Creativity Support Systems, pp. 192–205. University of Economics in Katowice, Katowice (2013)
15. Trąbka, J.: Organizational structure as a key perspective supporting the modeling of workflow processes - concept update and analytical tools. In: Proceedings of the International Conference on ICT Management for Global Competitiveness and Economic Growth in Emerging Economies (ICTM 2016), pp. 197–210. University of Wrocław, Wrocław (2016)
16. Tyrväinen, P., Päivärinta, T., Salminen, A., Iivari, J.: Characterizing the evolving research on enterprise content management. Eur. J. Inf. Syst. 15(6), 627–634 (2006)
17. vom Brocke, J., Seidel, S., Simons, A.: Bridging the gap between enterprise content management and creativity: a research framework. In: Proceedings of the 43rd Hawaii International Conference on System Sciences, Hawaii (2010)
18. vom Brocke, J., Simons, A., Cleven, A.: Towards a business process-oriented approach to enterprise content management: the ECM-blueprinting framework. Inf. Syst. e-Bus. Manag. 9, 475–496 (2011)
19. Wrycza, S.: Analiza i projektowanie systemów informatycznych zarządzania. Metodyki, techniki, narzędzia. Wydawnictwo Naukowe PWN, Warszawa (1999)
20. Yin, R.: Case Study Research: Design and Methods, 3rd edn. Sage, Thousand Oaks (2003)
21. Yourdon, E.: Modern Structured Analysis. Yourdon Press, Englewood Cliffs (1989)

Author Index

Printed in the United States
By Bookmasters